Into Africa

Craig Packer

INTO AFRICA

The University of Chicago Press / Chicago and London

Craig Packer is professor in the Department of Ecology, Evolution, and Behavior at the University of Minnesota. He has studied lions in the Serengeti and the Ngorongoro Crater and primates at Gombe since the 1970s.

The University of Chicago Press, Chicago 60637
The University of Chicago Press, Ltd., London

Printed in the United States of America
03 02 01 00 99 98 97 96 95 94 1 2 3 4 5
ISBN: 0-226-64429-4 (cloth)

Library of Congress Cataloging-in-Publication Data

Packer, Craig.
Into Africa / Craig Packer.
p. cm.
1. Packer, Craig—Journeys—Africa, East. 2. Ethologists—United States—Diaries. 3. Mammals—Behavior—Research—Africa, East. 4. Africa, East—Description and travel. I. Title.
QL31.P15A3 1994
599.051′09676—dc20
[B] 94-8428
CIP

This book is printed on acid-free paper.

Contents

PART III RED ROVER

SUDAN
ETHIOPIA
SOMALIA
ZAIRE
UGANDA
KENYA
Lake Victoria
Nairobi
RWANDA
Namanga
Serengeti
Mt. Kilimanjaro
Ngorongoro
BURUNDI
Arusha
Gombe
Kigoma
Tabora
TANZANIA
Lake Tanganyika
Dodoma
Dar es Salaam
N
East Africa
0
200 miles
0
300 kilometers
ZAMBIA
MOZAMBIQUE

Part I

Lion Eyes

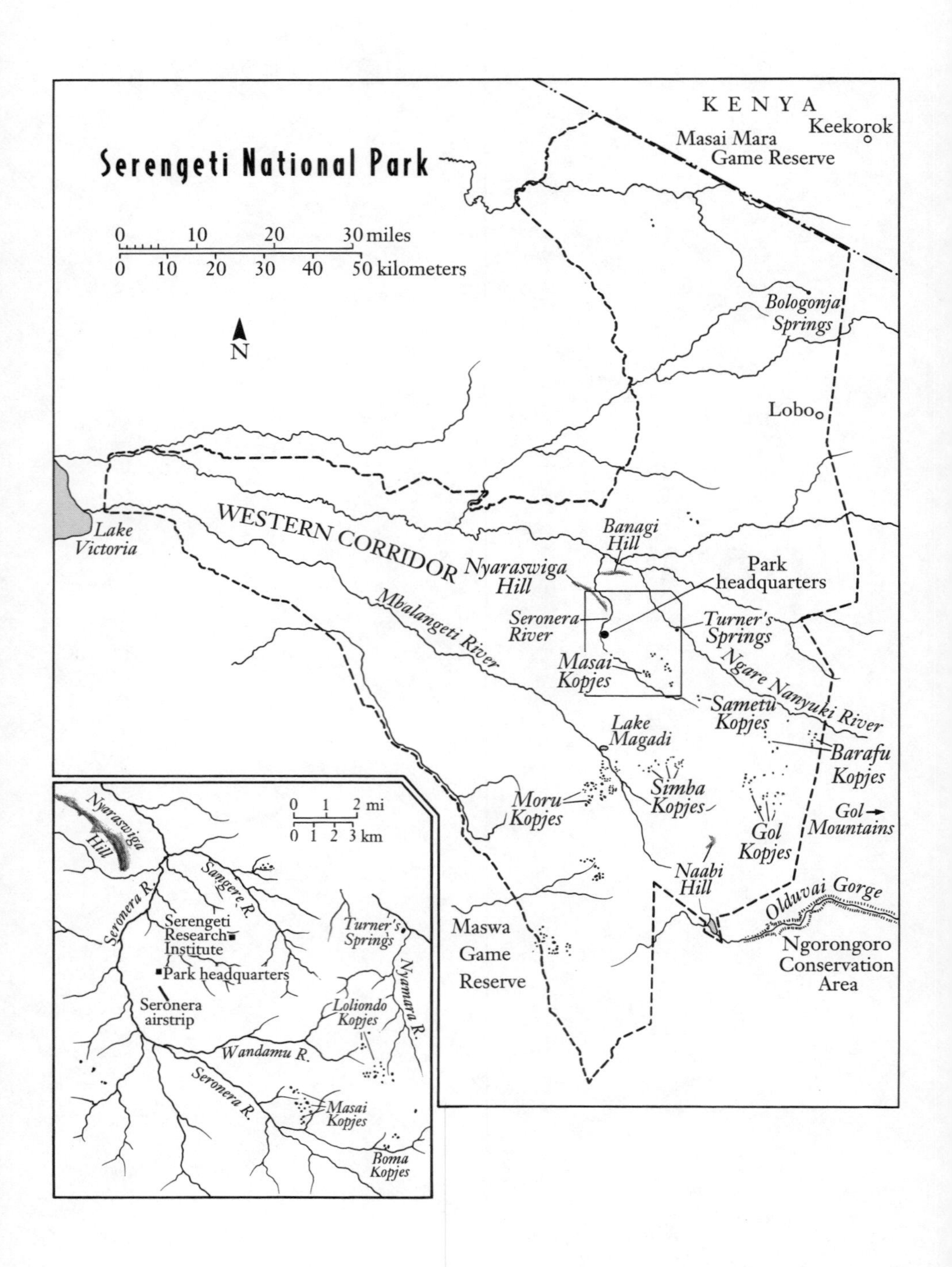
Serengeti National Park
0 10 20 30 miles
0 10 20 30 40 50 kilometers
N
KENYA
Keekorok
Masai Mara
Game Reserve
Bologonja
Springs
Lobo
Lake
Victoria
WESTERN CORRIDOR
Banagi
Hill
Nyaraswiga
Hill
Park
headquarters
Seronera
River
Turner's
Springs
Mbalangeti River
Masai
Kopjes
Ngare Nanyuki River
Sametu
Kopjes
Lake
Magadi
Barafu
Kopjes
Moru
Kopjes
Simba
Kopjes
Gol
Mountains
Gol
Kopjes
Naabi
Hill
Olduvai Gorge
Maswa
Game
Reserve
Ngorongoro
Conservation
Area
Nyaraswiga
Hill
0 1 2 mi
0 1 2 3 km
Sangere R.
Seronera R.
Serengeti
Research
Institute
Turner's
Springs
Park headquarters
Nyamara R.
Seronera
airstrip
Loliondo
Kopjes
Wandamu R.
Seronera R.
Masai
Kopjes
Boma
Kopjes

MINNEAPOLIS / SATURDAY, 26 OCTOBER 1991

In spite of myself, my heart is racing toward Africa. Behind my closed eyes I sense, for one fleeting moment, an impending warmth, glimpse a flash of brilliant colors. Green is mixed with gold, blue with black. I recall a world where time means more than the dull tick of anxiety, where individual existence is all but lost in the vast rhythms of life.

But the vision quickly passes and I am gripped by chronic exhaustion. Slumped in my seat, I am flying from Minneapolis to Nairobi via London and Rome. My ultimate destination is Tanzania, where I study the lions of the Serengeti and Ngorongoro Crater and collaborate with Jane Goodall on the baboons and chimpanzees of Gombe National Park. I am engrossed by the social evolution of these animals: Why do some individuals live longer than others? Why do some raise more babies than the rest? Why do they live in groups? Why do they cooperate?

Cooperation in nature is rare, but lions, baboons, and chimpanzees are among the most cooperative of all mammals. Lions are the most social of cats; chimps and baboons live in the most complex of primate societies. These are animals worthy of study in their own right, but their worlds also reveal something about ourselves. Human society may be the product of our own making, but I suspect we are motivated by many of the same desires and fears felt by any other animal.

This trip may sound exciting to most of the other passengers on board, but I am reluctant to leave the world of constant electricity and instant information. It is my sixteenth trip to Africa, and I have already traveled too much this year. I am a professor at the University of Minnesota. Most of my research has been done in collaboration with my wife, Anne, also a Minnesota professor. For over a decade, we spent half of each year in the Serengeti, but we can no longer shift our lives at the drop of a hat. Anne is staying behind in Minneapolis this time to catch up on her writing, and our kids need a chance to settle into school. They have already spent too much of their lives on the road.

So for the past few years we have spent as much time as possible investigating these animals on paper. The lions and baboons have been studied continuously for twenty-five years, and Jane has studied the chimps for thirty. Each of these long-term chronicles is equivalent to almost a century of human history. We have just finished cleaning up and collating the abstract biographies of all the lions, and the answers to many of our most important questions are now sitting in our computers, waiting to be revealed.

Now it is time to switch perspectives, to confront the living, breathing beasts. I would love to take a break from it all, but we are trying to solve fundamental problems about how cooperation can arise in an ocean of self-interest. The historical records must be kept unbroken, and so I return each year to the other side of the earth. The excitement of scientific discovery is still intense. Those few crystalline hours are worth any kind of hardship, any amount of hard work.

LONDON / SUNDAY, 27 OCTOBER

I arrive at 6:00 AM and wait for the others to appear at Heathrow Terminal Two. Combating cramp and fatigue, I move about as much as I can, wandering slowly between the garishly lit newsstands and the orange signs at the Alitalia desk. By mid-morning, the plastic highlights have all faded against the grey walls, grey floors, grey sky.

The hallway has gradually filled, and the terminal looks like an ad for the United Nations, with people from all over the world waiting peacefully for their flights. Hostile nations on neutral ground: Arabs, Asians, Africans, Europeans. But this taxonomy is too crude. We live on a tribal planet where local identities mean much more than race. I can't distinguish Norwegian from Dane, Syrian from Lebanese, Tahitian from Fijian; I can't read the lines of regional hostilities on each face.

In the waiting area, an Asian woman (Is she Indian? Bengali? Pakistani?) keeps a close watch on her little boy. He is not quite two years old. He wanders a few yards from his mother and finds something on the floor—a discarded baggage label. Waving it about in triumph, he charges around the alcove, victorious in his first battle of the day. He starts to lose all self-restraint, runs into strangers, whirls on the floor. Mother quickly rises and brings him back into line. She tries to instill a little discipline, then distracts him with a toy. The child converts the toy into a gun. Dozens of fellow passengers are mortally wounded. Mother tries to ignore the carnage. Shot for a fifth time, I finally fire back with my index finger. Tomorrow's warrior doesn't know whether to laugh or cry. Confused, he suddenly wilts and hides behind his mother's sari.

At 10:00 AM, Karen McComb arrives from Cambridge to swap video equipment and say hello. She has recently finished a two-year stint with the lions in the Serengeti and will spend the next few months in front of a television monitor in her college office, watching tape-recorded lions and reliving her years in the tropics.

Over the next half hour my three traveling companions appear one by one from different parts of Britain: Christine from Oxford, Sarah from Aberdeen, and Pamela from Belfast.

Christine is an ecological parasitologist who will be investigating the intestinal outpourings of our study animals. Christine came from Germany to Oxford, where Anne and I spent our sabbaticals last year. Our field associates have already sent her dozens of fecal samples from the Serengeti and Gombe. By looking for worms and eggs in each specimen, she has already discovered that some animals are more heavily infected than others. She can't yet tell why, and her research will succeed only if she can detect a clear pattern in her data. Do animals that eat together share the same worms? Are they healthier in drier habitats? Can monkeys be infected by humans? Christine will accompany me throughout this trip, then park herself on a lab bench for the next two years. She has a mere seven weeks to see the animals behind the samples, seven tightly scheduled weeks in which nothing can go wrong.

I have met Pam and Sarah only once before, when they interviewed for a three-year job as field assistant on the lion project. They had recently graduated from Edinburgh University and had conducted research in Scotland and Siberia. Anne and I had planned to hire only one person, but we liked both Pam and Sarah so much that we couldn't choose between them. It also seemed smarter to send them out as a team so that they could look after each other when confronted by equipment failures or by rapacious African officials. But we can't provide them with two decent salaries, so they will be dividing a single modest paycheck from our latest grant.

Our research is financed by the National Science Foundation, and we must anticipate our expenses with care. NSF supports most of the basic scientific research in the US, yet its annual budget only amounts to the cost of a single Stealth bomber. When requesting funds, we estimate the cost of the equipment, salaries, fuel, and car parts we will need over the next three years, and we try to gauge the severity of future inflation. If we make a mistake, we must live with the consequences. One of our research vehicles recently died and it is too late to request supplementary funding. Rugged, all-terrain vehicles are extremely expensive in the third world—especially with all the import duties imposed by the local governments. The new car has taken a huge bite out of our budget, and further expenditures will have to be kept to a minimum. Everyone will be working more for love than for money.

In this spirit of parsimony, we are flying today on ultra-cheap fares. But as soon as Pam drops her rucksack on the baggage scale, she realizes that she has misplaced her ticket to Nairobi. She is almost speechless with shock—she had organized everything so carefully. What else is missing? We hastily purchase a replacement ticket at the check-in counter, and the emergency one-way fare is the same as a discounted round-trip. However, the lost ticket can be refunded if it is eventually recovered.

On the bright side, the sympathetic Alitalia clerk checks our grossly overweight baggage through at no charge. Bag after bag passes over the scale, but she militantly refuses to read the dial. The overweight charges would easily have exceeded the cost of the lost airfare.

We all stand around in the crowded terminal, whirring with dizzy, half-witted airport excitement. Karen is vicariously elated by the others' imminent adventure; Christine and Sarah are both ready to fly with or without an airplane; only Pamela seems rooted to the ground, deeply embarrassed in her first day on the job. I feel myself starting to slip back into my African persona: slightly larger than life, up on my toes, ready to cope with anything, no matter how exotic the setting. Still reluctant to be going, but ready for action.

A three-hour stopover in Rome gives us enough time to try to call Pamela's flatmate to rescue her lost ticket. We need a polyglot who can deal with an Italian telephone operator. Neither Sarah, Pam, nor I share a common language with the man, but Christine knows a smattering of French. In the end, the attempt of a German speaking French to an Italian operator to use an American credit card to call Northern Ireland fails completely, but the effort passes the time and somehow keeps our spirits up. Pam and Sarah, too excited to sleep the night before, crash on the bright red Italian floor.

Another sleepless night lies ahead as we finally take off. Through the jet windows, the sparkling lights of Rome quickly pass from view. An inky black darkness engulfs us as we fly out over the Mediterranean. The north coast of Africa steals past undetected; the Dark Continent rolls by below. Occasional open fires glimmer weakly from the distant ground. The unseen void shifts from desert to savanna to forest and swamp; one major religion reluctantly yields to another. A dozen arbitrary boundaries encircle mortal enemies, and millions of competing viewpoints struggle to survive another night, another journey through the dark.

NAIROBI / MONDAY, 28 OCTOBER

The conveyor belt in the Jomo Kenyatta Airport delivers our twelfth piece of luggage. With four battered trolleys filled to overflowing, we look like a team of Victorian explorers setting out for the source of the Nile. The sign above the customs desk warns against smuggling seditious newspapers, books, or magazines into the country. The Kenyan press is heavily censored; international criticism is blocked at the border.

It is not yet dawn, and the customs officers are offhand and humorless. They can charge 250 percent duty on anything they please, and their curiosity seems to ebb and flow at random. But rather than let them take control, I start talking fast. "This is all research equipment. We will take everything down to Tanzania on Wednesday. Nothing will be left behind."

First I wave at Christine's large box of supplies marked "Scientific Equipment." Then I point to the liquid-nitrogen tank that will carry lion blood samples back to the US. The tank is labeled "AIDS Research, National Cancer Institute." No one ever asks to look inside the tank; they would rather be handed a venomous snake. The customs men wave us through without a murmur.

For the next few days, Pam and Sarah will be staying at Sarah's uncle's house; he has lived in Nairobi for over twenty years. Sarah's aunt greets us at the airport with the news that Pam's ticket has been found. Pam's burden is lifted.

Christine and I are greeted by my guardian angel, Barbie Allen. Barbie's father settled in Kenya following World War I, and she has lived here all her life. While her four children were being educated abroad in the late '70s, Barbie enlarged her family to include several itinerant scientists working in Kenya and Tanzania. Anne and I first met her in 1978, shortly after we moved to the Serengeti. Since then, Barbie has organized the lion project from afar, sending us emergency car parts and food, and even helping to arrange our kids' birthday parties.

From the Serengeti it is an eight-hour drive either to Nairobi or to Arusha, Tanzania, our other chief source of supplies. However, Kenya and Tanzania have followed very different economic policies since they gained independence from the British in the early 1960s. Kenya has been gung ho capitalist, while Tanzania has adopted a policy of idealistic socialism. These disparate policies led to a predictable divergence in living standards. Anne and I always relied on regular access to the luxuries of Nairobi, but because of the sheer difficulty of such long drives on miserable roads, we only came to town once every couple of months. Thus the help and supplies we have received from Barbie over the years have been worth their weight in gold.

Driving into town in the clear morning sun, we pass rows of flame trees, their flowers incandescent against the cool green leaves. The highway is bordered by hedges of bougainvillea, blossoming purple and orange. Jacaranda trees arch over the side streets and cover the pavement with a lavender blanket of fallen petals. We pass through the center of Nairobi, where traffic barely crawls and the air is clogged with exhaust fumes. The skyline has become ever more crowded with scaffolding and skyscrapers.

Conversation is lively, and I have to battle fatigue to keep up with Barbie's razor wit. She briefs me on the gossip of the past year, making my reassimilation more or less complete. My second life reclaims the front of my mind. People I haven't thought about since I was last out here suddenly seem very important, their affairs a matter of considerable interest. A whole cast of characters and their complex set of relationships replace the now-blurred crowd of family, friends, and colleagues that I knew so well somewhere else just a few days ago. It is almost as if I had never left Africa, never had another life anywhere else.

Barbie's split-level house is situated on the grounds of a small farm in an affluent suburb. We turn off the road and drive slowly past horses and stables, scattering the flock of geese that spends each day milling about the driveway. Emerging from the car, I greet Wahomey the driver, Odongo the houseboy, and Esther the cook. Some years I start out speaking Spanish or Japanese, but this year the Swahili is there when I need it.

We march unsteadily into the dining room for a lavish breakfast and meet several friends from the Serengeti over cereal and bacon and the thickest cream in the world. Revived for the moment, I start trying to cope with the Kenyan phone system—busy signals and unanswered phones are not always what they seem—then search around for the equipment I stored before my last departure.

Walking outside, I squint at the brilliant sunshine and try not to unwind too completely in the warmth of the African day. The well-practiced logistics of organizing our safari and shopping for the requirements of seven weeks impose a certain mental clarity that will last for only about six hours before I collapse in an exhausted heap.

In the late afternoon Christine and I go into town to look for a scale we can use to weigh the chimpanzees at Gombe. We start in the touristy part of Nairobi, near Woolworth's and the New Stanley Hotel. Alibhai Sharif's

Hardware Store carries a large selection of scales, but nothing that could withstand month after month of heavy rain. The salesman suggests a half dozen other shops, all in the more rundown part of town.

We step outside into the stagnant city air. The rains are late this year, and the afternoon is unusually hot and smoggy. Most of Nairobi's gleaming high-rises could belong to any Western city, but the newest reflect a wave from the East: brightly colored banks of Oman, Saudi Arabia, and Pakistan with narrow windows and high, vaulted arches.

The manic excitement of the journey has finally worn off, and we drag ourselves along the cluttered streets just as all the offices close for the day. Thousands of Africans stream out to join the already large crowds of pedestrians. Kenya has one of the fastest-growing populations in the world, and Nairobi doubles in size every ten years. Most vehicles run on diesel, and the car fumes trapped in the human flood create a choking, burning haze.

We walk as fast as we can through the rising tide. Streets become narrower as we enter an older part of town; excess humanity spills from the sidewalks into the traffic. My exhaustion suddenly gets the best of me, and I am terrified that the flowing crowd will separate me from Christine. She has never been to Nairobi before—doesn't know her way back to Barbie's—and it will be dark in another hour or so. I dart from curb to curb, wedging between cars and passing through countless pedestrians, frantically looking back with each maneuver to make sure that I haven't lost her. But we both stand out a mile against the background of African faces; there is no real risk of losing Christine in this crowd.

There are extraordinary responsibilities involved in bringing graduate students to Africa. They are expected to become independent scientists. They have to be protected from the hazards without becoming too reliant; they need room to grow, but you mustn't let them fail. Science is always a gamble. Some people win and others lose. You try to help your students down the narrow path between the impossible and the trivial, but you never know what's going to happen next. If you did, it wouldn't be worth doing.

TUESDAY, 29 OCTOBER

I wake up today feeling much more bleary than yesterday—must have been too tired to sleep, as my cousin from Texas always said. According to the local newspaper, the Minnesota Twins have won the sixth game of the World Series; the seventh game must have been played by now. But news of the outside world has become strangely irrelevant. The harsh sun bleaches out events in peripheral parts of the world.

Christine and I collect Pam and Sarah and spend most of the morning buying dry goods and household supplies, stalking the shelves of a supermarket on Kijabe Street. We load our carts with large bags of flour and rice, a dozen bottles of cooking oil, whole cases of fruit squash. Matches, mosquito coils, batteries, and don't forget to buy a large carton of toilet paper. Cookies, oatmeal, cereal, raisins, and cashews. Coffee and tea, honey, spices, sugar, and powdered milk. If you want it, get it now; no one's going to see a shop like this in Tanzania.

Just before lunch, we register Pam, Sarah, and Christine with the Flying Doctors, who will fly us out of the most far-flung spots in case of a medical emergency. Five years ago, Anne caught malaria in the Serengeti, and the Flying Doctors evacuated her to Nairobi Hospital. Malaria is the principal threat to our health out here, and the best protection is to avoid being bitten in the first place, so we buy four mosquito nets and a bottle of insecticide.

I finally purchase a scale to weigh the chimps, but it won't last long—it has no protective cover to keep out the rain, and everything at Gombe rots eventually. Christine finds the supplies she'll need to preserve hundreds of fecal samples: vast numbers of tubes and bottles and large quantities of formalin. Meanwhile, Pam and Sarah buy towels, bedding, and other essentials for their new home.

Returning to Barbie's in the late afternoon, we drive past the Finnish ambassador's house and the homes of wealthy Asian and European businessmen. An army of African night watchmen have started heading toward their posts. Most are dressed in the blue uniforms of Securicor; the rest are decked out in the grey and red of Ultimate Security. Crime is rampant in these wealthy neighborhoods. Gangs armed with machetes (known locally as *pangas*) head out for rich pickings from the squalid shantytowns on the outskirts of Nairobi. Over there is the house where a *panga* gang broke in and killed everyone before taking a few small appliances. Every hundred yards or so is a mast with a siren. Graceful living carries its costs.

At Barbie's house, Pam and Sarah finally see the long-wheelbase vehicle that will be their mobile research station for the next three years. It is just back from the garage where it has been repaired and serviced. Spare parts are virtually unobtainable in Tanzania, and most of our trips to Nairobi are made on behalf of our cars.

The Land Rover is bright red with white lettering on either side that reads "Serengeti Lion Project." It is a tall, square car, and the interior eventually will be refitted with a bed in the back so that Pam and Sarah can camp out next to the lions. It is now configured to carry passengers and supplies, ready to leave tomorrow.

In the evening we pack a dozen ox hearts in Barbie's deep freeze. These will serve as bait to administer antiparasitic drugs to the lions. Christine can learn much more about lion parasites if she can recover adult worms as well as their eggs, but I am unwilling to immobilize the lions just to inject them with worm medicine. We had been stumped for a good delivery system until Barbie suggested the hearts, which are cheap and just the right size to toss out the car window. Once frozen, they will fill our picnic cooler to the brim, leaving only enough room for a few small packets of frozen fish for the rest of us.

WEDNESDAY, 30 OCTOBER

With everyone and everything crunched into the red Land Rover, we wave good-bye to Barbie Allen, and head off: four adults, the worldly possessions of two young women moving to Africa for three years, the endless empty jars of an eager parasitologist, and a seven-week supply of provisions.

We bounce along the narrow drive from Barbie's peaceful haven onto a quiet suburban lane. The traffic becomes heavier and heavier as we head toward the center of Nairobi. We crawl past the high-rises, past the clenched fist, twenty feet high, that celebrates the authority of the single-party state. We rise over the railway bridge, above the tracks laid down by Asian coolies who were terrorized by man-eating lions in the 1890s, then pass the football stadium and approach the edge of town.

Whenever we hit a pothole, the cardboard boxes piled high in the back of the car come crashing down on Pam and Sarah. At a Caltex station just outside town, everyone concentrates on tying rucksacks and boxes together so they'll stay put for the rest of the journey. The service station crew fills our tank and checks our oil as a stealthy attendant lifts Christine's binoculars from between the two front seats. The theft remains unnoticed until we are an hour or so down the road. Though infuriated and affronted, we are behind schedule and must press on to avoid traveling these lonely roads at night. Gangs of thieves sometimes throw boxes of carpet tacks on the highway at dusk, then ambush any travelers who stop to mend the punctures.

We drive on through dry acacia brushland punctuated occasionally by small, one-horse towns that cater to the local Masai. We pass a small herd of Thomson's gazelle feeding on the edge of the pavement. The rain has fallen only sporadically here so far this year, and the runoff has nourished a narrow strip of green growth along either side of the road.

Masailand, stretching for another hundred miles to the south, follows the contours of the arid savanna that runs midway between the humid east coast and the rain-drenched heart of the continent. But the savanna was carved up by the English and the Germans during the nineteenth century. Kenya and Tanganyika were capriciously demarcated by the same process that converted an intricate web of hard-won tribal boundaries into the four dozen patchwork quilts that are now known as African nations. Some tribes remained intact within these new configurations, but others, such as the Masai, were chopped in two. Looming off to our left is perhaps the finest symbol of this imperial map-building whimsy: Queen Victoria redrew the Kenyan border to give Mount Kilimanjaro to her German grandson for his twenty-first birthday.

Up ahead, nestled at the foot of a stark, granite-peaked hill at Namanga, the Kenya/Tanzania borderpost looks at first like any other small African town. But the Kenyan side is a Mecca of opulence for the commerce-starved Tanzanians. Street vendors crowd the roadside, ready to hand socialist bus passengers their first taste of materialism: gaily colored plastic buckets, large nylon scrub brushes, and squat tins of cooking fat. Withered Masai grannies crowd around Western tourists, hoping to sell aboriginal trinkets to the palefaces. On the Tanzanian side stand a few blank buildings and even fewer signs of life.

Shortly after independence, Tanganyika merged with Zanzibar to become Tanzania. Aided by socialist advisers from the leftist days of post-war Britain, Tanzania established a set of economic policies with almost the same disregard for reality as Queen Victoria. After a dozen unhappy years united with Kenya and Uganda in the East African Community, Tanzania closed its border with free-market Kenya, and for seven years Namanga was as impenetrable as the Iron Curtain. The border finally reopened in 1985, but the disparity between the two economic systems still divides the two ends of town.

The immigration offices are low, featureless buildings shaded by broad verandahs. Dozens of aimless bystanders lounge in the shadows. We are prepared to charm our way past any belligerent official who feels like hassling us for transporting so much gear. But everyone is calm and distracted today, and neither customs, immigration, nor police take any interest in us on either side of the border. We pass through after only an hour.

Back in the Land Rover, Pam dispenses snacks as we drive along: banana chips, arrowroot crisps, and Indian spicy-nutty-stuff that spills all over the place and tastes great. The sun's rays blast through the open car windows, scorching knees and elbows that had faded white beneath the thick wrappings of winter. In the heat of the afternoon, the strong clear odor of rosemary permeates the car; Sarah has brought cuttings from her uncle's house to transplant in the Serengeti.

The countryside becomes drier and dustier. Parched open flatlands are surrounded by towering volcanoes. Snowcapped Kilimanjaro, like a monument to the moon, floats suspended in the mid-afternoon haze. Straight ahead is Mount Meru, a freestanding giant whose volcanic symmetry has been ruined by the ancient collapse of half its rim. The sun turns golden in the late afternoon, then inflates into a dull orange balloon before sailing forever west. We hit the fertile green highlands of Meru's southern flank and roll into the outskirts of Arusha just as night begins to fall.

The Tanzanian National Parks Service and the Wildlife Research Institute both have their headquarters in Arusha, and I must introduce each new lion worker when she or he first arrives. Although I used to dread each trip to Arusha, Tanzanian bureaucracy has become much more cooperative since the end of the Cold War. Westerners are no longer despised by lofty socialists attempting to preserve the purity of the masses, and decadent capitalists have become the only expatriate game in town. The Tanzanian government has recently adopted a more flexible economic policy and bowed to the demands of the International Monetary Fund. Even such shantytown nightspots as the Mother Cuba Bar and Grill and the Revolutionary Guest House have renamed themselves the Texas Bar and the Happytime Hotel. The country's entrepreneurs have been quick to keep up with global politics.

We spend the night in a posh hotel because I need a functional phone system to arrange for the trips to Dar es Salaam and Gombe. The hotel also has a walk-in freezer where we can store our hearts.

The four of us sit down for our first face-to-face dinner. The hotel has new French management, and the food is supposed to be very good, but the restaurant is almost empty. Three or four African waiters hover as we wait for the food to arrive. One of them finally assumes that Sarah doesn't know what to do in a restaurant: he quietly takes her unopened napkin from the table and spreads it on her lap.

The food lives up to its reputation and conversation ranges from the upheavals in Eastern Europe to the prospects of a united European community to why everyone but me is vegetarian. None of us seems to remember that we are supposed to have jet lag.

ARUSHA / THURSDAY, 31 OCTOBER

In the morning we meet David Babu, Director of Tanzania National Parks, who served as chief park warden in the Serengeti for many years. During Tanzania's socialist heyday, David suffered enormous frustration and indignity. The government devoted few resources to the parks system and discouraged donors from capitalist countries. The few precious spare parts that were imported for the antipoaching vehicles would disappear in the state warehouses of the capital city. Staff salaries would fail to arrive for months on end. The low point came when the park ran completely out of fuel, and operations ground to a halt. David couldn't even commute between his house and his office.

But he persevered and was widely regarded as the best warden in Tanzania when he was appointed parks director in 1985. David is an imposing man who demands absolute discipline from his staff, but he has had to lead under difficult conditions, operating within a system for which he has no respect but to which he has had to remain utterly loyal. This conflict often makes his interactions with foreigners quite awkward, and I never know how strongly I should agree with him when he recounts the problems that are dragging him down.

Today David is relaxed and gracious when we enter his modest office to pay our respects, and he warmly welcomes the three newcomers. He slips into his public persona as he tells us how things are going. Almost automatically, he recites the litany of familiar disasters: poaching threatens every park in the country, the roads are dreadful, there never seems to be enough money. But then he modulates into a more positive key. The central government is starting to treat these problems more seriously, international donors are increasing their support, the country's economy is steadily improving. There is finally reason to be optimistic. Tanzania is in the midst of a profound change, and because the government has acknowledged its past mistakes, David can be refreshingly honest about his public duties. But he is careful not to tell us anything about himself, about how he survives on an official salary of seventy dollars a month in a country where a bottle of beer costs five.

Our next stop is the headquarters of the Wildlife Research Institute to meet with George Sabuni. George was working on his master's degree when Anne and I first came to the Serengeti in 1978, and he has since completed his Ph.D. at Syracuse University in New York State. He came back to Tanzania a year ago to become WRI's third research coordinator.

George is still learning his way around as coordinator, but he has no illusions about the difficulties that lie ahead. Tanzania is a poor country with virtually no resources to spare for wildlife research, and WRI has had a sordid history. Among past directors and coordinators have been one outright thief who fled to America and a corrupt veterinarian who was last seen wandering the streets of Arusha, seedy and unshaven.

But George has returned from America with a more professional attitude. He has brought fresh ideas for raising money, made new efforts to promote cooperation with outside organizations, and, very much to my surprise, is holding our clearance from the Tanzania Research Council. With official research clearance we can receive residence permits and remain in the country for a full year. In the past, our clearances have often been delayed for months on end.

We sit down around George's large, cluttered desk and fill out the residence permit application forms as George translates the legalistic Swahili: exactly where we will be working, who our contacts will be within the country, the details from our passports. Once finished, we rush back to the middle of Arusha and escape from the intense noontime sun into the cool, open lobby of the immigration office.

Immigration officers hold the power of life and death over our dreams and aspirations in this country. They can confiscate our car and send us on the next bus to Kenya if they don't like the looks of us, and even when they do decide to cooperate, they can drag out their procedures for weeks.

On our best behavior, we knock on the door of room 7. A gruff woman in paramilitary uniform orders us inside and stares coldly at our applications. Beside her desk is an ancient rotary-dial telephone with a coiled black cord so hopelessly tangled that the receiver couldn't possibly be lifted off the cradle. The walls are lined with rough-hewn wooden shelves jammed with hundreds of dog-eared brown folders stacked awkwardly on their sides. On her desk lies the green Tanzanian passport of Miss Gladness Kasaya, the resident permit forms of Mr. Akbar Sheikh Yunus and Vipin Virji Velji. Scattered on the bare concrete floor are scraps of carbon paper and the torn visa application of Ruth Joel Tibazindwa, peasant.

The immigration officer looks up from our papers at last and announces that everyone's application is in order except mine. I have held previous permits and need to talk to the chief officer. Smiling as loudly as I can, I charge upstairs to room 11 and interrupt the reveries of the great man himself. The gold stars on the shoulders of his navy-blue uniform sparkle in the dull light. He wants to know why I keep coming back to Tanzania, and I explain that I want to know everything there is to know about lions, and they live a long time. This, clearly, is the least plausible story that he has heard in months, but he admires my audacity and instructs Miss Lonelyhearts to renew my permit. Instead, she sends us off to the center of town to get additional passport photos.

We walk down a side street in central Arusha where the pavement is crowded with pedestrians. African men wear faded Western shirts and dull trousers, but the women are a sight to behold. Indian women are clad in bright orange saris or yellow silk stove-pipe trousers, Pakistani women wear powder-blue veils, African women are wrapped in startlingly loud colors. One woman wears a pink-and-white blouse with a purple-and-yellow sweater and a polka-dot skirt of turquoise and orange. Then, incongruously, we see the occasional white face, harried and overly tanned, balding or wrinkled, consumed with impatience and hurry, burning out against the placid background.

The photography studio is in a small knickknack shop run by Asians. Here you can buy pictures of Indian gurus, cheap jewelry, assorted trinkets, and small toys. Mostly unobtainable until recently, the tawdry merchandise seems irresistible to the endless stream of African shoppers. We wait in the midst of a small crowd clustered around the single counter. The photographer is just coming, just coming. Puzzled shoppers gape and stare, but no one ever seems to buy anything.

Across the street, a small kiosk topped with a red and white sign advertises Tanzanian cigarettes: "TOP CLUB for men whose decisions are final." Various beggars and street merchants patrol the sidewalks. Walk by and you are accosted by old, half-naked Masai women thrusting beaded necklaces in your face. Sleazy men sell maps and homemade knives. Ignore their merchandise and the men all whisper, "Change money."

The black market thrives in Tanzania. Exchange rates on the parallel market are more than twice as favorable as the legal rates for tourists. But most change-money men are thieves. I've heard stories of street dealers who run away as soon as they get their hands on the hard currency or who stuff envelopes with squares of newspaper to mimic the feel of cash.

While my companions carry our photographs back to the immigration office, I head off to pay the final installment on our new research vehicle, the giant hole in our budgetary pocket. My postdoctoral associate on the lion project, Rob Heinsohn, started the transaction several months ago and drove back to the Serengeti yesterday in the new Toyota Land Cruiser. The Toyota dealership finally handed over the car when Rob convinced them that I would arrive today with the rest of the payment. To finish the deal, I have brought over nine thousand dollars to pay the taxes and duties. I have to pay in cash because the Tanzanian banks lack the technology to accept telegraphic transfer of funds.

The Toyota dealership is not permitted to accept foreign currency, and the bank is painfully slow. Each receipt must be signed by three different officials. Messengers carry the paperwork reluctantly from desk to desk. Dozens of people crowd around the counters, peering in at the tellers. My name is finally called, and two tellers hand over stack after stack of Tanzanian shillings. The largest denomination is a five-hundred-shilling note—the equivalent of a two-dollar bill. The money reeks of incense; only Asian shopkeepers handle such large sums of cash.

Having safely come this far with the dollars, I must now carry two million shillings in a large bag down the mean streets of Arusha. Surely every desperate ruffian and scoundrel can guess what I am carrying so awkwardly along the street; it is hard to be nonchalant walking past the ragged beggars, the dangerous and the dispossessed. No longer protected by my cordon sanitaire of young women, I have also attracted the attentions of the prostitutes who have already started working the streets, lifting their shiny nylon skirts and thrusting out their chests.

I walk faster and faster, past the London Bazaar, Kassam Traders, Ruby Accessories and Guest House, past sidewalk checker games of Fanta bottle caps played against Sprite. If I pause for even a moment, those whispered sleazy voices resume their insistent chant, "Change money." My anxiety increases until I can only see straight in front of me; time consists of the steady drip, drip, drip of sweat. I finally feel the door handle of the Toyota dealership and virtually leap inside, no longer crushed by the weight of all that hopeless, lawless poverty outside. The next half hour is an air-conditioned anticlimax, watching the Toyota sales department count to two million, five hundred shillings at a time.

Reunited with residence permits and a few thousand shillings left over from the Toyota transaction, we reach the Air Tanzania office just before closing time. Christine and I buy tickets to Dar es Salaam and Gombe and finally sort out the precise itinerary for the rest of our trip. After ten days in the Serengeti, we will fly from Arusha to Dar and then on to Gombe for ten days. We will return to Arusha at the end of November to finish with the lions and to attend a scientific conference on the Serengeti ecosystem.

Back at the hotel, I sit by the bed looking out at Mount Meru and prepare to spend all evening with the telephone. The lines are usually down over much of the country, and most phones only work a few days each month. But today seems to be my lucky day. First, I reach Jane Goodall's house in Dar to tell her assistant when we will arrive to organize and ship out all of Jane's long-term data on the chimps and baboons. Second, I ask Jane's agents in Kigoma to relay our schedule to the researchers at Gombe. Jane's research director, Anthony Collins, will meet our flight at the Kigoma airport and take us up to Gombe on the same day.

We go downstairs for supper at the hotel restaurant. The television in the bar is showing a video, and we pause to watch for a moment. The darkened room is lit by a familiar view of the Serengeti: umbrella acacias line a tributary of the Seronera River, the grass on Nyaraswiga Hill is a brilliant green. Then the red Land Rover rolls by on screen.

I hastily shepherd everyone straight on into the dining room. It is the film about our lion research, and I have seen it too many times before.

FRIDAY, 1 NOVEMBER

Rain fell during the night, and the air sparkles this morning as we repack the car. The summit of Mount Meru, dusted with fresh snow, looms above us like a broken Mount Fuji. We tie everyone's suitcases and rucksacks onto the roof rack to make space inside for large quantities of fresh fruit and vegetables. We finish off a few last scraps of business in town, and arrive at Arusha market mid-morning. The market is a haven for thieves, so Christine guards the car while the rest of us walk inside.

Many years ago I stopped here with my parents before taking them to the Serengeti. I left my mother in the front seat of the car with the keys in the ignition. A stranger tried to engage her in conversation through the driver's window. She realized that he was watching something over her shoulder and quickly turned around to see an arm removing her handbag from the opposite side. She had the presence of mind to close both windows, remove the keys from the ignition, and then give chase, shouting "Thief!" Even though she was nearly sixty, she was still fleet of foot and started to catch up. The thief panicked and threw her bag under a large lorry parked beside the road. She got down on her hands and knees, crawled underneath to retrieve her bag, and then emerged only to find herself surrounded by a small crowd of bystanders demanding that she reward them for their assistance.

Holding our breath through the heavy stench that emanates from the butchers' stalls, we enter a vast courtyard stuffed with produce and shoulder-to-shoulder humanity. We buy several large shopping baskets, then pass between a narrow passage of spice stands where the air vibrates with clove, ginger, garlic, and coriander, to the open-air vegetable market shaded by a high tin roof. Even in the indirect light, the colors are sensational. Bright red tomatoes you've seen only in your dreams. Brilliant orange carrots, mangoes, and citrus. A rich variety of greens: cabbages, beans, zucchini, limes, lettuce, and bell peppers. A collage of eggplant, pineapple, potatoes, and cauliflower. Shallow baskets filled with dried beans, lentils, peas, and sesame seeds.

We quickly find an African helper who works for everyone in the market. I negotiate prices and try to guess how much four people will eat, how long everything will last. Our helper runs to each stall to find the juiciest fruit, the healthiest veg, the freshest eggs—he keeps a tally of the sales from each vendor. We fill two baskets in minutes, putting potatoes and onions on the bottom, limes, oranges, peppers, eggplant, and zucchini in the middle, carrots, passion fruit, and green beans next, and on the top, unripe tomatoes. These are the best tomatoes in the world, but we can never buy them ripe—they would be pureed by the rough road between here and the Serengeti.

The fresh food fills the back of the car, two passengers stuff themselves into the back seat, the trays of eggs are wedged in somewhere safe, and we're off. We drive past shantytowns, through coffee plantations, and back to the dry savanna. Mud huts dot the side of the road. Masai herdsmen tend their cattle in the middle distance. We roll over hills and valleys, the pink soil half hidden by dry yellow grass. New landscapes unfold as the great northern volcanoes disappear behind us in the enveloping haze.

The highway is good for the first fifty miles. It has only recently been repaved by the Italians after being devastated in 1979. The Tanzanian army traveled over this road when it went into Uganda to overthrow Idi Amin. The government went bankrupt in the effort, and the road was converted to a spindly black lacework of asphalt draped over an endless series of potholes.

Although the new two-lane highway is pleasantly smooth, we never drive faster than fifty miles per hour. The road follows a high bed and passes over various bridges and culverts. Our cars spend most of their time on rugged terrain in the middle of a national park. The steering linkage is often loose and the tires are prone to sudden punctures.

And now the music has started. Driving along without tapes or radio, something is almost always playing in my head, filling the void. Right now it's the Eagles:

> I've got a peaceful, easy feeling
> And I know you won't let me down.
> 'Cause I'm already standing
> On the ground.

The song starts off vividly enough, harmonies intact and probably even in the right key, but I've lost it after an hour or so. The melody has died, the rhythm has been replaced by some random cadence of the road, and the lyrics have deteriorated:

> Bloddy-oh woe-dody-oh.
> Wah-duh-daw wuh-daw.

The paved road ends shortly after we reach the floor of the Rift Valley and turn off toward Lake Manyara, almost a mirage in the shimmering distance. The sheer wall of the rift escarpment stands behind the alkaline lake like a purple ribbon. Above the cliffs, a line of volcanoes marks the course of continental drift, a series of towering scabs coagulated along the ruptured surface of the earth.

The Land Rover judders over the corrugations, jostling the luggage inside and on top of the car. Chalk-white dust pours in through every crack in the bodywork, and the loose steering feels as though we might swerve out of control at any moment.

Perhaps ten miles along the lousy road we suffer our first puncture, and after much rearranging of baskets, boxes, and bags, we manage to extract our high-lift jack from beneath the back seat. Everyone pitches in—round stones block each wheel, a line of wheel nuts decorates the hood, the truck-size spare pops into place. With a smooth division of labor, the whole process takes only a few minutes, but we are beginning to run a bit late if we want to make it to the Serengeti before nightfall.

Lake Manyara is located on the floor of the Gregory Rift, where the altitude is only three thousand feet and the air is very hot. The broad plain is dotted with countless white smokestacks of clay, fifteen feet high. The twisted cooling towers have been built by termites to vent hot air from their underground colonies. Above the escarpment ahead, the cool highlands are dominated by the broad shoulders of Ngorongoro Crater. Younger, intact volcanoes rise higher still. Most impressive is Oldoinyo Lengai, the Mountain of God, which was last due to erupt six years ago. On particularly hazy days in the Serengeti we blame Lengai for the urban pall, assuming that it has finally cleared its throat.

At the northern border of Lake Manyara National Park we stop at the vegetable market in Mto-wa-Mbu, the River of Mosquitoes. Malaria is rampant here; tourists stop only for a Coke. Many of the Africans are transients, undesirables who have been officially banished from the major cities. The market lacks any real character; the shops along the outer ring sell uninspired wooden carvings and imported cotton cloth printed with ersatz African designs. But we quickly complete our safari menu from the wide selection of bananas and papayas at the ramshackle heart of the market.

It is now after two in the afternoon and very hot. We pass by the entrance to Manyara Park and start our ascent of the escarpment, winding past huge baobab trees, doubling back along dusty hairpin turns, and rapidly gaining enough height to see the groundwater forest below. Springwater percolates down from the Crater highlands, and clear streams gush from the cliffs of Manyara year-round. The permanent water sustains the lush forest on the valley floor, and the perennial green foliage supports the park's elephants, impala, and baboons. From halfway up the escarpment, we can see that the city limits of Mto-wa-Mbu form a visible boundary all along the northern edge of the park. Town and park have subdivided the oasis in the otherwise parched floor of the Rift Valley, and it is no surprise that Manyara's animals are so often in conflict with the neighboring townspeople. Manyara is a virtual breeding ground for baboons, and animals from the park frequently become banana thieves, thriving on the plantations at the edge of town. Periodically, whole baboon troops are trapped and shipped off to Western countries for biomedical research. The lions from the park also make the local villagers very nervous—many years ago, a Manyara lion specialized in catching drunks that reeled out of the town bars late at night.

Near the top of the first plateau, our engine is seriously overheating. Driving cross-country in the Serengeti, our cars harvest large quantities of grass seeds that become jammed in the vents around the radiator. The matted mass is a trap for dust, and soon there is virtually no air circulation to keep the engine cool.

Finally picking up speed to thirty-five miles per hour, we reach level ground and the engine temperature begins to stabilize. The soil up here is bright red and Ngorongoro Crater dominates the scene ahead and to the right. Onward and upward we go, into the cooler regions, past eucalyptus groves and horrible erosion from disastrous agricultural policies. Whole fields have become red, creased, barren. The topsoil that is the lifeblood of the highland farmers is draining down to the arid floor of the Rift Valley, where it threatens to silt up the shallow waters of Lake Manyara.

We stop in Karatu, the last town before the protected areas. Our radiator is boiling madly, and the worst climb is still ahead. The Karatu petrol station has no running water, but the attendant leads me to a cattle trough a hundred yards away, and we use a half-crushed margarine tin to transfer, drop by drop, the last inch of water from the trough into a rusty bucket.

By the time we reach the boundary of the Ngorongoro Conservation Area, the radiator is boiling again. We stop at the gate to sign in and refill a second time. This is only the midpoint between Arusha and the Serengeti, but at least we have reached the protected lands. The NCA was originally part of the Serengeti National Park, but the Crater highlands were reclassified in order to grant the Masai access to their traditional grazing lands. Although indigenous people are not permitted to live in the national parks, pastoralist Masai can dwell in the conservation area as long as they do not engage in intensive agriculture. The NCA protects large tracts of highland forest that serve as a water catchment for the entire region.

Tall, white-barked trees line the road, their upper branches draped with Spanish moss, the understory covered with choking vines. We grind up the steep mountainside past scenes of innumerable tragedies, sites where buses, lorries, and passenger cars all fled the winding road when their brakes, steering, or drivers failed. Over there is the spot where three Swiss missionaries were killed, up here a bus went over, and still higher is the bright yellow lorry that crashed last year.

Finally we reach the summit and our first chance to look down into the Crater. The upper slopes are richly verdant thanks to the fog brought in daily by the moist easterly winds. Up ahead, the rim is only as wide as the road. Off to the left, we take one last glance toward Karatu and the Rift

Valley; to the right lies the implausible circular symmetry of Ngorongoro Crater. But we don't slow down to enjoy the view. Thieves once attacked and murdered a tourist at the scenic turnout just ahead, and farther along is the spot where frustrated robbers killed another tourist when the truck carrying the wages for the NCA staff failed to appear one Thursday afternoon. Organized for a heist, and with nothing better to do, the bandits ambushed the last car of the day.

We surge past the tourist lodges perched atop the Crater rim, then on past NCA headquarters. The road is all downhill from here, and our radiator isn't so furious now that the afternoon has started to cool off. But then our wheels refuse to roll in a straight line. Another puncture.

Using our last spare, we admit defeat and head back to the Ngorongoro petrol station to waste the rest of the afternoon.

NGORONGORO / SATURDAY, 2 NOVEMBER

Checking out of the government-owned lodge seems to take forever. I recite a long series of complaints about the facilities, the service, and the prices, but the receptionists have heard it all too many times before. Their response hovers somewhere between embarrassment and indifference. But it's not their fault, they have no say in how the government runs this place, and besides, tourists come here for the view down into the Crater, not for the ambiance.

The stopover has meant a long time in the sunbaked car for the fresh vegetables, but Sarah's bundle of rosemary overwhelms the odors from this ripening cornucopia. Opening the car is like landing in Provence. A good thing, too, because we have failed to keep the fish and ox hearts frozen since Nairobi. The lodge freezer was as erratic as everything else, and the picnic cooler is best held at arm's length.

With a cool radiator and two new spare tires, we roll past dozens of Masai who are trying to attract our attention. Until recently, the Ngorongoro Masai had always maintained an aloof dignity, using the tourist roads only as cattle trails or to look for a lift. Today, though, groups of women stand along the roadside, hopping up and down, bobbing their necklaces and shouting "*Piga picha* (Take our picture)." Several of the women are dressed in electric blue rather than the traditional Masai red. At the top of the road that leads into the Crater, a small group of Masai warriors poses against the two-thousand-foot drop, silhouetted against the mirror lake, lush swamps, and impossibly sheer walls. In their gaudily feathered outfits they look like phony extras in a true-life, wide-screen spectacle.

Determined to get to the Serengeti before noon, we continue straight past the descent road and snake through the remaining Crater highlands in brilliant morning air. The open grasslands of the western slopes are dotted with scrubby, long-thorned acacias. Beyond a low bluff the landscape suddenly drops away to reveal the Siringit, the immense open space.

The Serengeti is framed to the north by worn volcanic highlands, the Gol mountains. Immediately to the south stands a wall of younger peaks, Lemagrut and Oldeani, still proud and dominant. That long gash down below is Olduvai Gorge, where discoveries of fossil hominids have redefined notions of human ancestry. And to the horizon, in front and beyond the curvature of the earth, lie the plains, the endless plains of the Serengeti. After the accessibly small scale of the Crater, the view ahead is symphonic—larger, so much larger, that it is only fully visible to the imagination, a vast intangible space teeming with life, bursting with experience.

The Serengeti is a giant clockwork driven by rainfall. The very heights upon which we are standing block the moist, easterly winds off the Indian Ocean for most of the year, leaving the western slopes in a barren rainfall shadow. Farther to the north and west, closer to Lake Victoria, rain falls during most months of the year, but the soil there is leached of vital nutrients. When the intertropical convergence passes through East Africa each year, the monsoon rains intensify until they can finally overwhelm any terrestrial barrier. Millions of wildebeest, zebra, and gazelle suddenly migrate in dense congregations to the rich grasslands of these volcanic plains; the soil here is fresh and the grasses are nutritious. Last night, we could see signs of distant rain, but today it is still dry, dusty, and empty down below.

That wide open space may seem serene when viewed from the air, but we are very much on the ground. The road demands the driver's full attention, and the brake pedal is the only nagging therapy I can offer in response to the car's headlong, precipitous deathwish down the steep, winding roadway. I grant the car a moment of peace when at last we reach the bottom, but then I start badgering it once more with the accelerator. Down here the level of dust suddenly thickens, and the corrugations in the road assume legendary proportions. This is the highway that kills cars by shaking them to bits. After a few painful miles, our carefully packed baskets have been rearranged by the forces of chaos, and we stop to tie everything back in place.

The road bumps, dusts, and addles us; conversation is impossible. We slowly crawl through the stair-step strata of Olduvai Gorge, past cryptic potholes filled with volcanic dust. If we hit one, the rest of the world disappears in a dense, billowing cloud; if we slow down, the trailing dust plume catches up with us in the following wind. Our hair assumes a strange dry, crackling-sponge quality; everyone has the same-colored clothes, same-colored skin. Our noses are clogged, our sense of smell has been vanquished. The car skitters along the washboard road, finding its own way, lurching from one shoulder to the next.

Upon reaching the superior roads of the Serengeti National Park, the car responds once more to my directions. Now I can look around and contemplate the emptiness of the plains. Dark clouds line the western horizon, but the rains have not yet reached this part of the park. Here all is yellow and brown, and the grass is as short as a fairway. To the west, the land is monotonously flat; to the east, the vague, rolling prairie recedes into nothingness. Straight ahead, Naabi Hill sits atop the stark landscape like a dark green limpet.

Naabi is a tree-covered island that shelters a few browsing antelope and provides refuge for our southernmost lion pride. The Naabi pride clings to existence at the very margin of the lions' habitable range. No pride of lions can sustain itself between here and Olduvai, so deep in the rain shadow of the Crater highlands. The relatively brief rainy spells attract migratory ungulates to Naabi for only a few months each year, but the soil dries rapidly, and no large herbivores can remain. The current lion pride at Naabi is the third to eke out an existence on the hill in the past twenty-five years. The first two prides failed to raise enough cubs and died out.

Naabi is also the site of the official entrance to the Serengeti National Park. I walk uncertainly from the car to the park office to sign in, my legs not quite numb but not quite mine anymore either. The rangers at the gate start to tell me about a sick male lion that has been hanging around Naabi for the past few months. He has virtually lost the use of one leg and the rangers have shot several grazing animals to feed him. Should they shoot something else for him?

"Where is he now?" I ask.

"Over there somewhere," they say.

But it turns out that no one has seen him for several weeks. It is very dry here now, and there is virtually no prey around the hill. The lion probably limped off somewhere far away. Anyway, it is bad policy to feed him; you don't want an injured lion to start associating people with food. At least twenty people live in this small island in the middle of the plains. There is no sense endangering the rangers' families or the rest of the Naabi pride.

We charge down the hill toward the middle of the park. Familiar landmarks come into view, hills and rivers with rolling, spacious names: Oldoinyo Olbaiye, Nyaraboro, Mbalangeti, Loyangalani, and straight ahead, with only its summit poking above the horizon, Nyaraswiga. Nyaraswiga is one of the tallest hills in the Serengeti and it dominates the view from the research institute.

We are still on the open plains, but leaving the rain shadow; the grass is already knee-high with a few scattered trees. We pass through our first set of kopjes, huge grey boulders strewn about by some distracted deity playing dice. Lions and leopards are the principal tenants here. The rocks provide excellent lookout points, and the trees and crevices provide shelter for cubs. All eyes search for our first lion, but no one seems to be at home.

Rain fell here yesterday and the road is still muddy. I try not to splash through the puddles because our windscreen wipers stopped working after we left Nairobi. So did our fuel gauge, emergency brake, turn indicators, and back door lock. The fuel tank has sprung a leak. This is average attrition from a single pass over this road.

Our spirits rise as we enter greener pastures and see the first herds of wildebeest and zebra. The animals are facing east, poised to head out onto the open plains should the rains soon pick up. We finally reach the Seronera River, which divides the plains from the woodlands. Though we are still sixty miles away from Lake Victoria, the river marks a wall of humidity that supports a diversity of animals: monkeys, insectivores, browsers, creatures that stay hidden in tall grass and thick brush.

We pass the Seronera tourist lodge on the left, bump over the well-traveled intersection to the park headquarters on the right, then continue through three more miles of trees and brush to the Serengeti Research Institute, to our house—the Lion House. A three-bedroom breeze-block house with large windows, high ceilings, and a peeling white roof, it is shaded by umbrella acacia trees and surrounded by huge rainbarrels, various tin sheds, and cross-country vehicles in varying states of delapidation.

Built by Texas A&M University to Southern suburban standards in 1972, our house was originally equipped with electricity from a generator half a mile away and running water piped seventy miles from the Kenyan border. But that was in the glory days, when the Tanzanian government actively supported wildlife research, and funds poured in from abroad. Once the Tanzanians hardened their ideological stance, these modern conveniences began to disappear. The research institute had essentially collapsed by the time we arrived in 1978, but we were prepared to raise our self-sufficiency to the maximum. I fitted the roof with solar panels and installed new wiring in every room, coated the asbestos roof with thick white paint, and asked Barbie Allen to organize a shipment of rainbarrels from Nairobi.

After thieves broke into the storeroom of the house, we walled up the original garage with concrete blocks, installed a heavy steel door, reinforced the roof, and invited all potential criminals from the staff village to see for themselves that there would be no further opportunities here. They looked around glumly, shook their heads, and one finally said "*Namna benki* (It's like a bank)." We built a new three-car garage to shade our heat-sensitive research equipment and added a fuel store in case Tanzania once again forgot to pay its foreign fuel bills. We don't much relish the idea of having to walk to work.

Rob and his friend Erin have their CD player turned up so loud they don't hear us step through the open glass doors from the verandah. The philodendrons are still climbing the high white walls of the living room, the potted plants are still flourishing on the windowsills. Rob and Erin enter from separate rooms. Rob is quietly excited—he flies to Australia in five days for a job interview. Erin, though, is very tense. She has taken a year off from college to be here with Rob and will return home in January. She is the youngest person at the research institute and the only one without a college degree. She has been helping Rob with his research and collecting lion feces for Christine, but she doesn't much like the work and resents the fact that I haven't given her a salary.

After greetings and introductions, we unload the car quickly to protect everything from the sun and the hyraxes, distributing the goods between two houses. Christine and I will be based here in the Lion House, and Pam and Sarah will stay next door in filmmaker Alan Root's unoccupied house. The four of us will dine together at Alan's to stay out of Rob and Erin's hair. Rob has been working for me for over a year, and the Lion House is really his house now. I have to get used to the fact that I'm just a short-term visitor.

By late afternoon all the vegetables have been washed and stored, the hearts and fish are in the freezer, suitcases have been unpacked, beds are made, and we have all had showers. I have found a spare pair of binoculars for Christine and finished talking with Rob about the most recent whereabouts of the local lions. We convert the red Land Rover from transport vehicle to animal tracker by reattaching a directional antenna to the top of the roof rack and reconnecting the radio receiver.

Pam, Sarah, and Christine join me in the car once more, and we head back toward park headquarters. I put on a pair of headphones and explain how the radiotelemetry system works.

"We have radio collars on about two dozen lions, mostly in far-flung or brushy parts of the park. Each collar is made from a wide band of machine belting. A small radio transmitter is riveted to the belting and hangs loosely around the lion's neck. Each transmitter broadcasts at a different frequency, just like radio stations distributed across the dial. Since most of the lions restrict their movements to a particular area, you will have a good idea which individuals you might find before you head out, so you'll only have to program a few frequencies into the receiver. The receiver tunes in to the frequency of one lion for a few seconds, then automatically changes stations to allow you to listen for the next lion, then the next.

"The signal is an incessant beep that pulses about once a second—the closer you get to the collar, the louder the beep. The antenna on the car roof is highly directional, and the beeps are loudest when you drive straight toward the collar. Once you hear something, stop the scanner, drive in a tight circle to get the direction, and drive more or less straight over to the lion.

"You will often track the lions to places where you won't be able to see them, but at least you'll know where they are keeping themselves. If they pull up stakes and wander out of their usual range, you'll have to track them from an airplane—the transmitter has only a three-mile range from the ground."

The drive along the Seronera River totally disorients the three newcomers. They are all watching too closely for lions—under the shade of those trees lining the river, maybe, or somewhere in the brush—and they soon lose track of the major features of the landscape.

At about 5:30 PM I pick up a signal, a female from the Campsite pride, and we quickly track her to a large palm grove, down a steep riverbank. Her signal is perfectly loud and clear, but she is hidden in impenetrable brush. She probably has a new set of cubs down there, and she will certainly remain hidden until well after dark.

Everyone begins to realize the difficulties and frustrations that lie ahead. I double back along the river and drive into the heart of the tourist area where none of the lions are fitted with collars. The sun has vanished behind a low bank of cloud, and the evening light is dull and grey. I stop frequently and scan with binoculars. No luck. We rejoin the Arusha road and head back home for the night.

Pam and Sarah are not especially bothered by our failure to see any lions. They will be here for the next three years and are distracted by the physical reality of making a new home for themselves. They're just as happy to go back and unpack. Christine, on the other hand, wants everything to go one-two-three. She has just a few weeks, and each tick of the tropical clock is a big chunk of her time in Wonderland.

For me, it's just another evening drive along the same familiar road, past the same landmarks on the same old roadways. I can still recognize each tree and termite mound. Home again.

SERENGETI / SUNDAY, 3 NOVEMBER

Pam and Sarah's kitchen is spacious, but the floor is cluttered with half-emptied boxes and baskets. The early morning light shines in through the windows and bounces off the snowy white cabinets and countertops. The cabinet above the stove has been scorched by some ancient flambé, the white linoleum curls up slightly from the counter, and a fine layer of dust covers everything. The plumbing no longer works, and a blue bucket of rainwater stands in the middle of the floor. The cupboard shelves are filled with colorful plastic bowls, plates, and cups. The kettle has started whistling on the enamel stove.

"Where are the fixings for the cereal?"

"Raisins and cashews are in the larder, the rest is on the shelf by the back door."

"Which fruit needs to be eaten first?"

"That smashed papaya is still all right at one end."

"Who wants coffee?"

Then the inevitable queue to the loo before assembling at the car.

"Have you got your binoculars? Specimen jars?"

We finally set off at eight. The morning air is already warm. We drive along the SRI airstrip past Moses Rock, named after a young male lion that was attacked and killed there one day by a neighboring pride. At a nearby hyena den, only a few females are still out in the open; the rest have sought shelter from the intensifying sun. We follow a wall of tall trees that burst from a narrow riverbed like a runaway hedgerow. Upright fever trees sprout between squat sausage trees. Arriving at a broad ridge top, we drive, radio-tracking, in a loop around the Masai kopjes and listen for beeps in every direction. Not a sound.

Down toward the Seronera River, two lions are heading for the shade of a large, yellow-barked acacia tree. A female walks in the lead, low and golden, and an adult male follows attentively, his dark brown mane proudly proclaiming his manhood.

A lion pride is a tightly knit group of adult females, their dependent young, and a coalition of immigrant males. The pride territory belongs to the females, who pass it on from mother to daughter for generations. The males come into the pride as a group. They father the cubs and defend the pride against marauding bands of wandering males. Every few years the resident coalition is replaced by yet another group of males. Husbands come and husbands go, but the matriarchy carries on forever.

Each of our study prides is named after a geographical feature prominent in its territory, and each lion has its own identification card. Opening the card file, I look up the six females of the Masai Kopje pride, turning almost immediately to a female named MSL. She was born in the Masai pride in 1981 and has a handful of unmistakable markings. The male trailing a few yards behind her should be one of the three residents: John, Maynard, or Smith.

We name each lion the first time we see it. Each newborn cub receives a series of initials or numbers—MSL was the first cub born in the MS (Masai) pride after MSJ and MSK. Although this system may sound confusing, it is important that each animal's name be unique. Female lions typically give birth to litters of two to four cubs, and a pride may produce dozens of cubs each year. Cub mortality is very high, and a lot of good names have been used up over the past twenty-five years. Rather than struggle to find new names that may well be lost within a few weeks, we give each cub the next letter or number on the list.

However, we do give proper names to nomads. Male lions leave their pride of birth between the ages of two and four, and some males wander for years. Our study area only covers about a fourth of the Serengeti, and males come here from all parts of the park. By the time they reach us, they are usually quite distinctive in appearance, and inspiration comes easily. John, Maynard, and Smith, for example, first showed up as subadults shortly after a visit to SRI by the British biologist John Maynard Smith. Their limp manes were about the same length as his.

After inspecting this male's face through binoculars, I quickly sketch his whisker spots and ears. Each lion has four or five parallel rows of vibrissae, or whisker spots, on either side of its muzzle. We concentrate on the few extra spots above the topmost row (the "reference row"). Sometimes the spots are in the shape of a triangle. Other lions have just two or three spots in a straight line. The spots usually differ on each side of the face, and their relative position above the reference row is different on every animal. Whisker spots are easily detected in even the smallest cubs, and they remain the same throughout the lion's life.

From frequent spats with his companions over kills, this male has acquired a clear set of ear notches, converting his left ear into a cogwheel. Comparing my sketch with the ID cards of the three resident males, we quickly confirm that he is Maynard. After wandering the plains for a year or two, Maynard and his companions entered the Masai Kopje pride three years ago. Now nearly nine, Maynard is a bit past his prime; males are lucky if they live to be as old as twelve. Maynard's mane is still long and full, and his physique would be the envy of any weight lifter, but his jowls have started to droop, and his teeth are slightly yellowed. The triangular tip of his nose is no longer pink, but flecked black with tiny scars.

He lies beside the female like a somewhat twitchy sphinx. MSL must be sexually cycling. At ten, she is still in her peak breeding years. Like most female mammals, female lions live longer than males. Serengeti females can live as long as eighteen years and remain fertile until they are about fifteen.

The day is already hot, and MSL is ready to nap, but Maynard remains alert, not wanting to lose her to one of his companions. Lions are one of the few species in nature in which males form close, lifelong bonds with other males. Sets of young males born in the same pride will remain together throughout their lives, taking over new prides together and guarding their prize against other coalitions. Large male coalitions can always outcompete smaller ones and are much more successful in the genetic sweepstakes. Coalitions of six or seven males may father over forty surviving offspring.

Within the coalition, however, it is every lion for himself, and one male usually monopolizes a female throughout her receptive period. The consorting male stays within a few feet of the female for four or five days, mating with her several times an hour. He becomes very aggressive and warns his companions not to come too close to his lady. His coalition partners respect his powerful teeth and claws and do not challenge his rights to the female. As a result, overt fighting between partners is rare, and only the consorting male mates with the female.

But this priority of access depends on who gets there first. Should Maynard fall asleep and MSL wander off, he would lose all rights of ownership. Similarly, if MSL is not impregnated during this cycle, John or Smith may be the male of the moment in another two weeks.

About fifteen years ago, the original John Maynard Smith developed a simple mathematical model that could account for the evolution of such respect for ownership. He showed that when fighting is very costly, an individual that fights only when it is the closest to the resource (that is, when it is the "owner") will suffer far fewer injuries than individuals that fight at every opportunity. In a world full of mindless brutes that are capable of inflicting injuries with costs far greater than the value of their prize money, an individual that only fights for its own property will be more likely to live to fight another day, more likely to raise a family. The first individual to respect the rights of property will populate the next generation with an abundance of like-minded offspring. As the behavior spreads, fighting will become virtually nonexistent; once all individuals behave in the same way, rivals will never challenge owners and owners will never be provoked. Mutually assured destruction will result in a pattern of behavior that mimics the rule of law.

The three new lionologists are happy to see their first lions, and amused to see ownership in a male named after the theory's progenitor. I am pleased to see Maynard because my previous field assistants have only managed to collect blood samples from John and Smith. The three males had been born outside our study area, and we have no way of guessing from their behavior or appearance if they are closely related. Instead, we must rely on DNA fingerprinting to measure their genetic relationship. Knowing whether coalition partners are close relatives is important to us because kinship provides an essential foundation for cooperative behavior in most animal species. However, male lions will often cooperate with unrelated companions. These three males figured prominently in the behavioral research project of one of my recent graduate students, and I want to resolve their kinship in time for him to include the answer in his thesis.

MSL is already in deep flop, and Maynard only occasionally opens one eye; they won't be going anywhere between now and this evening. We note our precise location on the map and look around for distinctive landmarks to help us find this particular tree when we come back to sample Maynard.

Continuing along the Seronera River into open grasslands, we see a smattering of wildebeest, zebra, and gazelle attracted by the recent rains. Some are venturing farther out onto the plains, but there has not yet been enough rain to stimulate much grass growth. The land is still dominated by last year's dead grass, dull yellow and pink beneath the hazy blue sky.

However, there has been enough rain to reduce our chances of finding any uncollared lions today. During dry weather, lions position themselves around the river courses in order to ambush thirsty prey. Since most visitors come to the Serengeti to look for the big cats, all of the tourist roads run parallel to the streams. Once it has rained, though, there is standing water everywhere, and the lions have to change their tactics. They hunt mostly at night and actively stalk their prey in the open, well away from the rivers and the roads. In this hot weather, the lions have already melted by now and could be flopped out anywhere, concealed in the faded grass.

We still can't pick up a signal from the local lion with the collar, so we continue along the upper reaches of the stream, past the Boma kopjes, then north toward the woodlands, winding through the Loliondo kopjes and the upper end of the Nyamara River. The receiver scans different frequencies as we pass through each pride's territory, but no one is within range.

I use the time to orient the three newcomers: "Which hill is that over there? Nyaraswiga. And there is Banagi. That one's Nyaroboro. Follow this track to the main Arusha road; that track goes directly back to SRI."

We leave all tracks behind and drive across country through open brushland; past scattered herds of wildebeest and gazelle. Up ahead, a lone adult male lion sits upon the ridge top. He is somewhat shy, so we approach him cautiously. He starts to avoid us when we drive to within fifty yards, but his leg is badly wounded, and he is barely able to walk. He is a transient, a nomad, and he appears to have been the victim of a gang attack by a hostile male coalition. He has deep bite wounds on the inside of his thigh, along his back, and around his testicles.

Coalitions of resident males discourage trespassing. Three or four males must have surrounded him and bitten him from behind whenever he turned to defend himself. Bald patches on either side of his neck show where his bright orange mane has recently fallen out. The wandering hero has lost his mantle.

We drive up slowly until we can make out his whisker spots and ear notches. He sits awkwardly, nervous of this strange red object filled with unknown creatures that keep staring at him through long glass eyes. But he is not sufficiently frightened to rise up in pain and walk away. Through our binoculars we see that he has a large notch in one ear and clear spots on both sides of his muzzle. His appearance is quite distinctive, and from the color of his nose and teeth I guess that he is about six years old. An animal this shy has probably grown up in a distant part of the park. The Serengeti covers a vast expanse, most of which is inaccessible to visitors. The number of nomadic males we encounter each year indicates how well the lions are faring in these far-flung areas where poaching and trophy hunting are rampant.

We have never seen this male before, so he needs a name. Pam quickly christens him Cassidy. Cassidy is too crippled to hunt for himself, and without health or companions he will never gain residence in a pride. No other species would dare attack a lion, and he will endure long months in pain before finally starving to death. For most lions death is neither quick nor noble, and death comes earliest to those without allies.

Leaving Cassidy, we continue farther to the north. From a prominent ridge top I pick up the distant signal of the collared female from the Loliondo pride, L23. The radio signal leads us deeper into the woodlands, up to the banks of the Nyamara River. We follow the stream until we are opposite Turner's Springs kopje. The riverbanks are relatively low, and a submerged kopje provides a solid bottom. The crossing here is passable, but Sarah has to get out and tell me where to direct the tires.

We are in real thorn-brush country now, and radio-tracking is much more tedious. Huge stands of young acacia trees, only six to ten feet tall, obscure the view and limit the route we can follow. I drive in a half circle to the left, now to the right, trying not to drive directly over too many thornbushes. Our tires are ten-ply Michelins, but they are not bulletproof.

We finally spot L23 from only thirty yards away, and she, too, is an object of desire. She is being courted by MS18, the eighteenth male born in the Masai pride after we had gone through the alphabet. Finding these two by themselves leaves me feeling out of touch with the local soap opera.

The Loliondo pride has only been seen a few times in the past two years. They have shifted their range to the wrong side of a nearly uncrossable river, and this is the first time they have been accessible in months. We can see that MS18's coalition is still resident here, but where is the rest of the pride? Are all eleven females still alive? Have they delivered a new batch of cubs?

It is late morning and time to get back home. We have to finish unpacking and prepare everything to immobilize Maynard. Heading directly back to SRI, we pass over a series of low wooded ridges. The landscape here is far greener than on the plains, and herd after herd of wildebeest and gazelle stand ruminating in the shade of mature *Acacia tortilis*, umbrella-shaped trees that must surely have inspired the first parasol.

At home, we sacrifice the worst victims of the trip from Arusha. What may once have been green tomatoes, eggplant, and zucchini are all merged into a lunchtime glop. I run over to the steel-belted storeroom ("the vault") at the Lion House, pull out our collection of cassettes, grab a large grey boombox, and wander back in time to provide a musical background to the vegetable stew on Pam and Sarah's verandah. The aroma of rosemary waltzes to Chopin as we stretch our legs and admire the view. Just to the west, the sensuous bare surface of an orange and grey kopje sweeps down to the center of the front yard. Along the east side of the house, a rough-hewn wall of boulders shelters mongooses, snakes, and hyraxes. A hundred yards in front of us runs a small river lined with bulbous green acacias, their canopies dotted with fluffy white flowers.

In the riverbed, a woodpecker taps on the sounding board of a weathered dead tree as a Bateleur eagle and a pale chanting goshawk perch on nearby branches. Orange and white impala and tiny grey dik-dik stand out against the green thicket surrounding the river; and on the horizon, Nyaraswiga and Banagi hills flank the exquisitely symmetrical Kubukubu, whose four rounded bumps combine to form the outline of a giant tortoise.

Three or four grey flycatchers perch on the backs of our chairs, waiting for us to toss them tidbits of cheese. Weaverbirds and starlings, sparrows and waxbills all stand on the ground in front of the verandah looking for crumbs. A blue-eared glossy starling cries plaintively from an overhanging tree, staring down at us with its unearthly blue plumage and piercing red eyes. On the ground, a Ruppel's long-tailed glossy starling struts through the timid flock of smaller birds with all the warmth and charm of an SS officer, claiming each crumb as its own. Canaries and barbets flit in and out. Mousebirds crash-land in the middle of the fidgeting congregation.

Hyraxes, like large, furry stomachs with teeth, graze in the nearby grass. Tailless and with a face midway between Alfred Hitchcock and Goofy, they are said to be the closest living relatives of the elephant, though they look more like mutant rodents. Whatever their evolutionary affinities, they have voracious appetites and eat the birds' food as well as houseplants and kitchen scraps. Pam and Sarah want to start a garden here, but the hyraxes would devour every succulent shoot.

It is hot in the sunshine, but our elevation is five thousand feet, and in the shade of the verandah the air is fresh, cool, and dry. Except for the harsh, glaring sunlight and exotic animals, we could be in California or somewhere along the Mediterranean.

Over at the Lion House, Rob is preparing for his job interview, and Erin is still frosty. She has collected over forty scat samples for Christine in the past four months, but her strident lack of personal enthusiasm makes it difficult to generate any warmth in the conversation.

Keeping myself at a safe distance on the verandah, I check that the dart gun is clean and in good order, reconstitute a bottle of anesthetic, fill the barrel of a dart, and attach the plastic fins. Christine prepares an ox heart to give Maynard a dose of medicine before darting him. Maynard may wake up feeling a bit woozy, but at least he will be wormless for a few weeks.

The sun is already low in the sky, and we need to get back to the Seronera River before the lions saunter off for the evening. About two hundred yards from the Lion House, the car engine suddenly gives off a harsh squeal, and the charge light flashes red on the control panel.

Cursing, I get out and look under the hood. The fan belt seems okay. Maybe the alternator has given out. We don't have time for a major car failure—not now or any time this week. On the principle that simple-minded machinery sometimes forgets that it is broken, I start the ignition again. No squeal, no charge light. Everything seems okay again.

We retrace our morning route, and from the Masai kopjes we see two tourist vehicles that have stopped next to MSL and Maynard's tree. That means the lions haven't moved anywhere yet, but we will have to wait until the tourists lose interest before moving in ourselves. On average, visitors spend only ten minutes with lions before moving off to look for more animate organisms. Lions are a big letdown to most of the people who have traveled thousands of miles to see them. They may be big and impressive, but they seldom move.

The tourists quickly fill their photographic quota and drive off. We arrive just in time to see Maynard walking determinedly away from MSL.

This is a relief, because I don't like to dart lions when they are preoccupied with sex. Maynard goes to the river to drink and, after a very deep draught, starts walking along the road. We drive cross-country to get in front of him, stop, and then toss the ox heart onto the roadway.

He walks straight past the heart without slowing down. But downwind he stops to sniff the air. He turns back a few paces, sniffs the ground near the heart, then loses interest and continues on as before.

After he has walked fifty or sixty yards, I hop out of the Land Rover and retrieve the broken heart. Most of the drug has already leaked out, but we have to try to tempt him again.

We drive past Maynard and toss the heart a second time. Again, he continues past without even slowing down. I fetch the now-mangled mess again; we will keep it handy for any other lions we encounter tomorrow. But now we have to get on with the darting before nightfall.

Maynard is a big target, but he keeps walking at a steady pace. We leapfrog past him several more times before he finally stops to listen to his distant companions who have started to roar. Maynard stands stock-still, ears cocked, looking away from us toward the north.

From only fifteen yards away he is as big as a barn door. I fire and hit him square on the rump. He jumps forward an inch, but then quickly regains his dignity and carries on walking as before. A few minutes later he stops once more to listen, and after a moment of uncertainty, lies down on his belly, still intent on the world around him. After another two minutes, he is obviously feeling the anesthetic and bobs slightly before resting his head on his paw. We wait another two minutes while treating the inside of a large syringe with anticoagulant and slipping on rubber gloves.

Keeping out of his field of view, I drive to within a few feet of him. His yellow eyes are still wide open, but he doesn't move. I open the door, get out of the car, and bend down to tickle his ear. If he is too lightly anesthetized, he will twitch it in obvious annoyance. Stroking the outer edge of his ear provokes only a vague response. He is ready. The others can step out of the car, and we can start to handle him. Now comes the hard part.

Lions' skin is very tough, and their blood vessels are well protected from the teeth and claws of their opponents. Some veterinarians I've known have no trouble finding a vein in a lion's wrist or even in the tail, but others can't do it at all. I can draw blood reliably only from the large vein on the inside of the thigh, right next to the femoral artery. There the skin is soft, and I can actually see where to aim the needle, but I can only succeed if someone holds the lion's back leg straight out from his body. Pam and Sarah grasp the appropriate ankle and pull, and the syringe is quickly filled.

We hurry to examine him, beginning with his teeth. One upper incisor is missing, one canine is slightly broken. His chest girth is 48 inches. His weight must therefore be about 350 pounds, fairly average for a Serengeti male. I inject him with one worm medicine while Christine squirts the other into his mouth. He can't move, but he hasn't lost his swallowing reflex.

When we have finished poking and measuring him, we stand back a few yards and just look at him in the dying light. He is very, very big and very solid. His paws are massive. His head is enormous. He is helpless now, but he will soon be up again, padding softly through the night. Not someone you would want to meet on foot.

Off in the distance, a male lion starts to roar. The still evening air is filled with a sequence of long, tense groans; uuuunnnnhhh, uuuunnnnhhh, uuuunnnnhhh, uuuunnnnhhh. Each of his groans is a melodic crescendo: low basso at the beginning, rising in intensity before descending in pitch at the end in a pattern that is distinctively his own. He follows the groans with a staccato series of deep monotone grunts: unh, unh, unh, unh, unh, unh, unh, unh, unh, unh. His finale is an exhausted, descending sigh: uuunh. The whole performance has lasted about thirty seconds.

Lions almost always roar at night, when it is cool. Roaring is hard work, and lions don't like to strain themselves in the heat of the day. Lions mostly roar in calm weather. They don't waste energy when it is too windy for their calls to be carried into the night. On still nights you can hear a lion roar from five miles away. Females roar as well as males, but their voices are not as deep or as loud. Lions roar most during seasons of plentiful prey, when nomads or neighbors might be tempted to pilfer from their larder. Roaring tells the strangers how many lions occupy a given territory and where they are at the moment. Roaring also tells companions what is happening within the pride.

Lions are intensely social animals and do almost everything as a group, and not just as a mob but often as a team. They hunt together, rear their young communally, and defend jointly held territories. Anne and I have spent over a decade investigating different aspects of their cooperative behavior, but lions spend most of their time doing absolutely nothing, and our research has often involved countless hours of anticipation and frustration. But the lions' roars provide an important key to their social lives, and even if the lions remain inert most of the time, we can use recorded roars to conjure up cooperation on demand.

Each sex is fiercely territorial. Males are on the constant lookout to keep other males away from "their" pride, and females are intolerant of any strange female that wanders into their territory. Should a large group encounter a smaller one, the larger chases the smaller group out and attacks and kills any lion they can catch—remember what happened to Cassidy. Lions become very excited when a stranger roars in the middle of their range.

Seven years ago, Anne and I hauled out a pile of high-powered sound equipment from America. We brought a hundred-watt car amplifier designed for audiophiles who like to drive convertibles with the tops down, and a speaker designed to broadcast at high volumes. After we recorded a number of roars, we broadcast the recordings to lions in neighboring prides. Anne parked next to the lions to observe their reactions, while I set out the loudspeaker about two hundred yards away.

Whenever we played the recorded roar of a nomadic male to a group of resident males, the residents instantly became alert, listening intently. Once the recording ended, they would quickly rise and walk determinedly toward the speaker, ready to evict the invader. On they came, continuing right past the speaker, within a few feet of the car, heading off in the direction where the stranger must be. Gunslingers ready for the last showdown. If I had parked below a ridge top or next to a kopje, the lions would carry on until they had a clear view behind me. Once they decided there was no stranger nearby, they would lie down, look off into the distance for a while, then finally go back to sleep.

Just to make sure the lions were really coming to attack the invader, I arranged to have a stuffed lion sent from an English museum. We hid the dummy behind a bush just beyond the speaker and broadcast a roar. As soon as the lions discovered the dummy, they stalked it, approaching cautiously from behind, inching forward ever so slowly, before suddenly leaping up and attacking it. After they had ripped out a mouthful of tanned hide and styrofoam they quickly lost interest and, I have to admit, seemed slightly embarrassed.

The exciting part of the whole ruse was that the recording inspired the entire group of resident males to come forward together in defense of their territory. It is one of the few opportunities anyone has ever had to study animal cooperation experimentally. Once we discovered that recorded roars could elicit such dramatic responses, we launched a full-scale program to carry on the research. We brought out Jon Grinnell from Minnesota and Karen McComb from Cambridge to study roaring full time. Jon focused on the males, and his research took advantage of a curious pattern in the kinship of male coalition partners.

In most animals, cooperation involves close kin: honeybee workers devote their lives to the service of their mothers and sisters, and in several bird species maturing offspring help their parents to rear younger siblings. The evolution of cooperation in these cases is not terribly surprising because such close relatives have at least half of their genes in common. In 1964, evolutionary biologist William D. Hamilton developed a simple rule to predict when generosity toward close kin could evolve by natural selection. To "pay" for a suicidal act of self-sacrifice, the altruist's action must confer on its companion an additional number of offspring that exceeds the inverse of the proportion of shared genes carried by the two individuals. Thus absolute self-sacrifice can be advantageous if an altruistic individual more than doubles the reproductive output of its parents or full siblings, but self-sacrifice between full cousins requires more than an eightfold increase in successful reproduction. Of course, self-sacrifice can be far less extreme than this, but the ratio of benefits to costs from any act of heroism will have to follow Hamilton's Rule, and kinship will generally have to be quite close to influence an animal's behavior.

Lion prides are family groups: the females are all mothers and daughters, sisters and cousins, aunts and nieces. Sons of these "sisterhoods" grow up together as a cohort, and if there are enough of them to take over a new pride together, the young males remain intact as a coalition of brothers and cousins. Such coalitions may include as many as nine males.

But not all male coalition partners are close relatives. If a pride produces only a small cohort of cubs, or if most members of the cohort are female, then a male may reach maturity without any closely related male companions. Once he is evicted from the pride of his birth, a lone male is unable to gain entry into a new pride and spends years as a solitary nomad. During this phase, most solitary males team up with unrelated partners to form coalitions of two or three companions before finally taking over a new pride. Coalition formation of this sort is so common that two-thirds of all male pairs and nearly half of the trios contain unrelated partners.

This kinship pattern had puzzled us for a long time. The very fact that lions live in prides and breed synchronously means that there will often be large groups of male relatives, and that solitary males will have to team up to be able to compete against them. The larger the coalition, the better able it is to take over a pride in the first place, and the longer it is able to stay in residence, fathering offspring all the while. But solitary males team up mostly to form pairs, more occasionally forming trios, and never forming a quartet. Since, on average, a male's mating success continues to increase when coalition size increases beyond three, why do solitary males form such small coalitions? Why don't seventy-five males band together and control all of the female lions from the Serengeti to Nairobi?

We decided that the answer must lie in the pattern of mating success *within* each coalition and spent over five years collecting blood samples in order to test for paternity. We had already collaborated on several genetics studies with Steve O'Brien, who heads a large lab at the National Cancer Institute in the U.S. O'Brien assigned one of his graduate students, Dennis Gilbert, to measure fatherhood in the lions by analyzing their DNA.

This technique, called DNA fingerprinting, had been used successfully to identify human rapists and murder victims and to establish paternity in birds as well as people. We just had to collect enough blood from each animal to obtain a sufficient quantity of white blood cells, each of which contains a tiny amount of DNA. Collecting white cells from lions in the Serengeti and shipping the samples safely to Maryland involved considerable logistical difficulties, but our hard work eventually paid off.

Dennis extracted hundreds of genetic fingerprints and discovered that males only achieve equal mating success in pairs. In the larger coalitions, competition between partners for mating opportunities is so intense that many males are left virtually childless. Membership in a large coalition, therefore, mostly benefits the few breeding males.

Cooperation between nonrelatives has to include a high probability that each male will receive a direct genetic benefit from his behavior. *Fraternité*, on the other hand, is not *egalité*. Relatives will continue to benefit from cooperation even if they are shamelessly exploited by their closest kin, as long as their behavior satisfies Hamilton's rule.

Dennis's discoveries explained why males refused to team up with large numbers of nonrelatives, but we still needed to know why nonrelatives cooperated at all. Cooperation between unrelated companions can most easily persist if it confers such great mutual benefit that neither individual is tempted to exploit the companion's goodwill. For example, two individuals may need to build a bridge to cross a river in order to obtain food from the other side. If two are far better than one at building the bridge, and there is little risk to either of being hurt while working on the project, then we would expect them both to participate happily. This is called mutualism, and it is a felicitous state of affairs that does not often arise in nature.

On the other hand, if two heads aren't much better than one, and if it is really quite hazardous to help out and, well, there does seem to be a fair amount of food on this side of the river today, one individual will be tempted to let the other guy do all the work.

This second scenario presents a dilemma. Both individuals would prosper if they each cooperated, but each would be most successful if he could get his companion to do all the work. In this case, cooperation between nonrelatives may evolve, but only when there are many opportunities for cooperation with the same partner, when each individual can recognize if his companion is shirking, and when, in response, he refuses to do anything else for the common good. This kind of cooperation is called reciprocity and is virtually unknown among animals.

So what is going on in these small coalitions of unrelated males? Are they in a divine state of mutualistic grace? Or are they tempted to leave all the dirty work of kicking out strangers to their companions and to keep the ladies for themselves?

It would seem that male lions are faced with a dilemma: getting up from the safety of their slumber and chasing out a large, well-armed invader must carry considerable risks to life and limb. Maybe cooperation depends on kinship after all, and males will only take the risk for their brothers.

Perhaps small coalitions do more poorly than larger ones precisely because they are often composed of nonrelatives who might not work so well together.

Hence Jon Grinnell's playback experiments. Over a period of nearly two years, Jon played recorded roars of one, two, and three strangers to a variety of lions, either related or unrelated. His results clearly indicate that unrelated males are just as cooperative as close relatives and that they show no inclination to cheat.

Even in the face of unfavorable odds (three recorded lions against two real lions), unrelated companions were just as prompt to arrive at the speaker as brothers. Nonrelatives also worked for the good of the pair when their companion was far away and out of view. If their cooperation is based on reciprocity, a male would cheat when by himself—I really would be better off in the long run if I didn't stick my neck out right now, my partner isn't watching me, and he can't blame me for something he can't see.

Instead, Jon's experiments beautifully illustrate the unconditional affection a male lion holds for his companion. We have followed the frightened lives of solitary males that have survived for years before finally finding suitable companions. Once they forge a partnership, these males are quite touching in their warmth for each other. They rub past each other like lovesick teenagers, showing an attachment that they seldom display to females. Once they have made the decision to team up, they are as close as brothers; you cannot determine the degree of their genetic kinship from their behavior.

Day-to-day behavior in these males is part of a long-term relationship guided by mutual dependency. It is so important to have a companion that a male lion does whatever he can to keep him, and that means cooperating with him and protecting him at every opportunity. Both companions have a similar interest in keeping rival coalitions away from their females. If a male doesn't help his buddy today, his buddy may well get nailed by one of his enemies, and the shirking male will end up having to do all the work by himself tomorrow. Even if he is not killed at the next encounter, he may be ousted, never to gain entrance into another pride.

Your buddy is your lifeline. Do something that causes him to be hurt in the short term, and you hurt yourself in the long term. All the corny

Hollywood movies about male bonding in wartime may be awful, but they are somehow believable. Our moral choices are often clearest in the face of imminent disaster.

John, Maynard, and Smith were frequent subjects of Jon's study, but they entered the study area as nomads, and we don't know their genealogical ties. To confirm the generality of Jon's results, we need to know the kinship of every coalition that he observed. In the course of his genetic studies, Dennis Gilbert discovered that close relatives have DNA fingerprints much more like each other's than like those of nonrelatives. That's why we are sitting in the dark with a syringe full of Maynard's blood, waiting for him to wake up again. Sometime in the next few months we will finally discover whether John and Smith are Maynard's brothers.

There is not much moon, and the night has become cloudy. Maynard is beginning to stir, and his partners have started roaring again; friends and family surround him on all sides. He lifts his head and rolls onto his chest; he is already alert enough to terrify any curious hyenas who might approach him, and he will be up on his feet again soon.

It is time for us to head back to SRI. Time to make supper and try to unwind.

MONDAY, 4 NOVEMBER

Awake before dawn, I lie beneath the mosquito net that enshrouds the walls of my bedroom. This used to be my daughter Catherine's room. For her first few years, she slept here in an old wooden crib that Barbie Allen found in Nairobi. After Jonathan was born, we moved the crib to his room, gave Catherine a real bed, and installed this gossamer contraption to provide a safe space whenever she needed to get up in the night to use her potty.

Outside, the birds are still silent, and nothing is stirring. The glistening stars cast just enough light to silhouette the flashlight and malaria medicine on my bedside table. I could use more sleep, but it is hard to turn off after so much traveling, and there is too much to do in the next few days. I need to census several prides, replace a few radio collars, arrange a radio-tracking flight with the pilot at Seronera. Giving up on sleep, I turn on a small fluorescent lamp and finish unpacking. In the dim light I could be anywhere. Beyond my translucent cocoon of mosquito netting, the whitewashed walls, hardwood cabinets, and casement windows belong to no particular time or place. Africa is kept outside; we are safely tucked into a world of our own making.

The morning sky begins to brighten. I walk quietly out to the storage vault to look for spare parts for the radio-tracking equipment. The car antenna didn't seem to be working properly yesterday; one of the connections may be loose, or the coaxial cable might be broken.

Unlocking the heavy steel door, I step inside and turn on another fluorescent light. The floor is cluttered with spare car parts, tools, solar batteries, and tanks of liquid nitrogen. The left wall is completely hidden behind a deep set of shelves piled high with carefully stored belongings and equipment. But among the black metal trunks, suitcases, and cardboard boxes, I find another piece of my past life here: a bright yellow teeter-totter and two small boxes labeled "Toys." Nearly three years have passed since my whole family has been to Africa. When Anne and I lived in the Serengeti, we didn't look upon this house as a research station but as our home, especially after the children were born.

Our nanny, Betty, also had two children, and they were Catherine's best friends. The kids cooked cakes in the sandbox, clambered over a wooden climbing frame, whooshed back and forth on the swing set, and played with toys on the verandah. Betty was from an agricultural tribe near Arusha, the Chagga, and her children taught Catherine a smattering of Swahili and passed on several strong opinions about members of other tribes. One game involved playing the part of a Masai baby—and being thoroughly scrubbed and de-liced by the other two kids.

"You be Masai baby."

"No. You be Masai baby."

The antenna spares are approximately where I left them last year, but other things seem out of whack. Coming back to where I lived for so long, it is hard to mark the passage of time. Things haven't really changed, or have they? Did I put the drill back in its proper place the last time I used it? Was that a few days ago, or a few years? Maybe Rob has it.

I always feel a bit absentminded when I first come back to the Serengeti. The rest of the year is so completely irrelevant to life out here that time seems to pass at twice the normal speed. Memory must have a half-life, decay is surely inevitable; I must be forgetting something. But I'll never know for sure.

The sun is up, and Christine and I rejoin Pam and Sarah for breakfast. We have tried to establish a faster morning routine: we mixed the cereal last night, our alarms have been set for half an hour earlier, the equipment is all ready to go. But for some reason none of this preparation translates into action. We still don't get off until nearly eight.

Our top priority this morning is to find Maynard. The worm medicine we administered last night causes the worms to loosen their grip on the intestinal walls. Maynard's first bowel movement this morning will be their last, and we haven't got much time before he disappears for the day.

No leisurely cross-country tour this morning; I whizbang along the main road from Seronera past the park headquarters, the petrol station, then the lodge—watch out for the impala that leap across the road in a transcendent moment of grace before bounding into the bush, their horns cocked back along their necks. Warthogs at the sides of the road continue grazing, down on their front knees. Pigs always eat with their elbows on the table.

We arrive at the spot where we left Maynard last night. No sign of lions anywhere. We drive slowly down the other side of the Seronera River, then up the nearby Wandamu. Nothing. On top of the low ridge tops on either side of the two rivers, I stop repeatedly to climb onto the roof of the car and scan with binoculars. My glasses are wedged between the broad rim of my hat and the eyepieces of my binoculars. Too many years in the tropics without a hat means that I can't go anywhere without one now; wages of past sins keep my dermatologist in whiskey, as she likes to say.

Still no sign of lions. We are better off cutting our losses and looking for a lion with a collar. Heading back toward the Masai kopjes, we pick up a faint beep. It is the signal of MKP, a subadult male from the Masai Kopje pride. He is with two other lions, almost out of view on top of the easternmost kopje. We can't identify the others from this distance, and if we were to drive any closer, they would be hidden behind a high wall of rock.

Life would be easier if we could just climb these rocks on foot and inspect the lions' whisker spots up close, but after millennia of persecution by well-armed humans, they have a deep-seated fear of people. Fortunately, they do not seem to recognize car passengers as human, but once they see someone stand clear of the car, they flee in terror. During the daytime, anyway.

Driving around the other side of the kopje, we find a flat, narrow passageway to the top. I put the car axle into low ratio and slowly clunk and grind over fist-sized rubble. Surrounded on three sides by trees and boulders, we drive to within fifteen yards of two young males. These are MKP's cohort-mates: MKQ and MKR. MKQ has started to sprout a decent mane, but MKR is still almost maneless. They look like scruffy adolescents, junior high school students who should have started shaving several months ago.

Both lions seem completely worn out from whatever they were up to last night, and they barely notice our loud mechanical arrival in the heart of their fortress. Christine loads a new heart with worm medicine and passes it to me. I slowly step outside, keeping myself behind the open car door, then toss the heart.

It splats down about three yards from MKR, who doesn't even lift his head. We wait while he snoozes, oblivious to us and our offering. Rejected again.

Perhaps we could drive over the chaotic rock surface, retrieve the heart, and toss it again. Maybe we should load up a second heart and try hitting someone in the face.

MKP suddenly appears from the other side of a ledge, sniffing. He keeps his nose down to the ground and walks slowly toward the heart when MKQ suddenly passes him, ducks in, snatches up the heart, and eats it, leaving MKP to sniff the ground.

The mood in the car has now lifted about ten feet. MKQ finishes his snack, sniffs around for another few seconds, and wanders out of view. We prepare a second heart and toss it toward MKP. It splits in half on impact. MKQ hears the splat and comes back almost immediately. By the time MKP gets down to the heart, MKQ is there too, and they end up with half a heart each.

We still have one more heart in the car, the one Maynard had ignored last night. But we don't know how much of the drug has leaked out, and we don't really want to give MKQ a third dose. We stick around long enough to see if either MKP or MKQ show any reaction to the medicine. They merely fall asleep, as all good lions do at this time of day.

The wind has started blowing, the weather has changed, and the ground is drying out rapidly. Small herds of wildebeest and zebra are moving off the plains toward greener pastures to the north and west. Dust clouds trail along behind as we drive to the park headquarters. We park the Land Rover under the shade of a huge umbrella acacia in front of the wardens' offices. In the radio room, a ranger is shouting a message to someone in Lake Manyara Park. He intersperses rapid bursts of Swahili with lilting English and spells out key words with a phonetic alphabet: "*Tutapeleka barua lakini sasa* check for spares *hakuna* fu-el fil-ters *au* con-den-ser. That's C-Charlie, O-Oscar, N-Nairobi, D-Delta, E-Echo, Nai-robi, Sugar, Echo, Robert."

The response from Manyara is hopelessly garbled, and the static is tremendous. The ranger resumes his shouting: "Negative. Negative. C-O-N-D-E-N Sugar, Echo, Robert." More noise in response, and this time it seems that the information has been passed on. In an emergency we could theoretically send a radio message to the outside world via the Tanzania National Parks office in Arusha, but it is not a system you would want to stake your life on.

Hats in hand, we greet the chief park warden's secretary.

"*Hujambo, mama. Habari yako?* (Hello, madame. How are you?) May we see the chief park warden?"

She replies without lifting her eyes from her ancient typewriter, "All of the wardens are out in the field today."

"Will he be here tomorrow?"

She keeps typing with her index fingers. "Yes, you will see him tomor-row, come back tomorrow."

"*Ahsante sana* (Thank you very much)."

We negotiate our way past the armed ranger who guards the gate to the Seronera garage. He stands behind the rickety wire frame, clutching the stock of an ancient rifle, trying to look formidable. He is meant to keep out undesirables who might distract the workers and pillage the stores, but he acts as if he is expecting a full-scale invasion.

The workshop area is surrounded by a high wall with broken glass embedded in the top. Inside is an extensive collection of exhausted vehicles: Land Rovers, Land Cruisers, Unimogs, and lorries with their fenders, wheels, or engines removed. Some have been worn out while protecting the Serengeti, others were crashed by drunken rangers hurrying back from the village bar. I have spent more time at this garage than I like to recall. We come here for repairs, to have punctures mended, and to buy diesel. The ground is black with fossil fuel, the mechanics are mostly dressed in rags, the sun is truly fierce. From a corner of the main shed comes a steady thumping sound. *Bwana Pancha* (Mr. Puncture) is taking a tire off its rim with an old axle shaft and a couple of tire irons.

Daniel, still the chief mechanic, proffers a wrist for me to grasp. His hands are smeared with grease, but he somehow manages to look immaculate. He is in the middle of a delicate operation on a Toyota Land Cruiser. Then there is Joseph, who drives the parks lorry once a week to Arusha or Mwanza. We once spent several days together when he guided me to a lion that had been snared out in the westernmost part of the park. It was a male that he often saw on his way to and from town, and he was extremely pleased when we darted the animal and removed the wire loop from around its neck.

And here's the old, old carpenter whose name I've never known; I just address him as *mzee* (old man), a title of honor and respect in a land where precious few survive to an advanced age. The *mzee* once came to the Lion House to mend a door that had been eaten by termites, and since then I have acknowledged him with great deference, greeting him with the humble "*Shikamoo* (I hold your feet)," to which he invariably, and with great satisfaction, replies, "*Marahaba* (Delighted)!"

The workshop is still cluttered with the same old broken-down machinery. Over in the corner is the rusty arc welder that Awazi once used to mend the broken chassis of our first Land Rover. He cut a series of metal plates with a blowtorch and arc-welded each piece with such speed that the car frame glowed red-hot. Awazi was top man at the time, and crowds always gathered to watch him perform. When he was sure that everyone was watching, he casually reached over and lit a cigarette from the glowing frame. One of the garage minions was so impressed that he immediately went over to a grinding wheel, turned it on, and pressed a narrow steel chisel down hard against the spinning stone, sending off a huge spray of sparks. He then used the glowing red tip to light his own cigarette.

Pam and Sarah may have been formally introduced and oriented, but they are still blissfully unaware of the many long hours they will spend here in the next few years, dealing with the aftershock of the roads from town and the sudden consequences of innumerable warthog holes, hyena dens, and rocks discovered unexpectedly during cross-country excursions.

At the Lion House we set up a large camping table, and Christine prepares the fecal specimens that Erin collected last night. She quietly extracts a thumb-sized chunk from each stinking black sample, drops it into a plastic tube filled with formalin, and writes the lion's name on the label. I escape to the kitchen, pull Maynard's blood out of the refrigerator, and take it to the small office that serves as the blood-processing lab. The whole procedure takes about an hour and a half: spinning off the serum in the centrifuge, separating the red cells from the white cells, and decanting the three components into tubes for storage in the liquid-nitrogen tank. The white cells contain the DNA for the kinship study, the red cells are rich in enzymes that indicate overall genetic variability, and the serum is full of antibodies for immunological analysis.

A few months ago Steve O'Brien's staff discovered that about ninety percent of our lions are seropositive for FIV (feline immune deficiency virus) and that many have been exposed to feline herpes, feline leukemia virus, and a host of other horrors. FIV is particularly interesting, of course, because of HIV, or human immunodeficiency virus, which causes AIDS. FIV is not hazardous to the lions—infected animals live as long as those that have not been infected—but learning how the lion's immune system copes with the virus may be important to AIDS research.

Rob is in the living room drafting the talk he will give for his upcoming job interview in Australia. While the centrifuge roars in the next room, we talk about how he should present himself to the search committee. Rob's Ph.D. thesis was on a highly sociable Australian bird, the white-winged chough, and he was specifically interested in how young choughs learn to forage for themselves. He joined the lion project late last year to find out how young lions learn to hunt large, difficult prey, and he has also been performing playbacks to subadults to study how they acquire the territorial behavior of their parents. He suffered a slow start because of chronic car trouble, but he is riding high with the new Toyota and optimistic about the months ahead.

By late afternoon Pam and Sarah have finished cleaning out their house and Christine has organized the parasitology lab to her satisfaction. Pam has been baking bread, and the kitchen is covered with baking tins and fresh loaves. Sarah has started making macramé pot hangers for her herb garden, but the hyraxes have already eaten most of her cuttings and the rosemary sprigs have all died. Planning for dinner, we inspect the fish from Nairobi. We can approach only one packet to within twenty feet; there will be one more vegetarian around here by tomorrow.

At six o'clock, we head back to the Masai kopjes. From the steady beeping of MKP's collar we know that they are still up in their stony castle. We sit in the dying light and wait for them to come into the open.

Just after the fading sky has stolen the last colors from the world, the three young males quietly emerge from the kopje and walk purposefully down to the road. They are almost invisible against the dead grass. From behind, we can make out the black bars on the backs of their ears and the random motion of their black tail tips. We strain to figure out who is who. Only two of them received medicine. MKP has the collar, so he's easy, and MKQ has a slightly larger mane than the others, but the night will soon be too dark to make such subtle distinctions.

The males walk about half a mile along the track, plodding slowly, slowly. Up ahead, a group of tourists is camping illegally by one of the Masai kopjes. The lions walk straight toward their camp, and we turn on our headlights to outline the oncoming trio to the milling campers, intent only on setting up their tents. One of the lions stops and stares at the people, much less bothered by them at night than he would be during the day.

Lions are almost a different species at night. This is their element; this is when they really do seem like kings, afraid of nothing, less concerned about being seen by either humans or their prey. Their night vision is much better than that of most other animals, and they seem to know it.

The three young males walk beyond the campers and fan out over a forty-yard front, stopping frequently to listen. Hyenas are up to something somewhere off to the north. If the hyenas have made a kill, our boys will dash down to claim it for themselves. We follow about thirty yards behind as the lions walk on sporadically for another half mile. But then they plop down in the grass, out for the count.

A few minutes later we hear the pounding hooves of wildebeest and the whinnying and snorting of zebra stallions. The riot grows louder, moving closer to us. Two of the lions get up and trot toward the herd, which seems to have been disturbed by something, possibly the hyenas. The lions stop and stare off into the distance.

The night is now pitch black; we occasionally flash our headlights to locate everyone. The herd is quite close and running toward us. In one flash of the lights, we see several wildebeest running desperately away from something in the darkness. One of the uncollared lions is standing only thirty yards away.

On the next flash, more wildebeest are coming, and the lion is moving toward them, making no effort to conceal himself—no stealth, no cunning, no strategy. Males are poor hunters even at the best of times, but these three seem to have no idea how to catch their own supper. Although nearly full-sized, they are barely two years old and will not be fully mature until they are about five.

The herd passes us by and the three males just stand around. It is difficult to say whether they were trying to catch anything or just diverting the wildebeest toward unseen hyenas. Hyenas are excellent hunters, and male lions often rely on them to act as their bird dogs, shooing the hyenas away like flies once they catch something.

All three of the young males are now lying down. Two are still alert, but the third is seriously asleep. They may already have expended their full quota of energy for the night.

My passengers are all excited, expecting constant action. Although lions do often show a burst of activity in the cool of the evening, they usually settle back down again and sleep most of the night. Lions have to be unpredictable to capture their prey, and they are perfectly happy to lie still and wait for dinner to come to them. They will occasionally interrupt their slumbers to sit up and look around, scanning the landscape in case something is wandering blindly toward them. But they never seem to be in any hurry. When they do finally bestir themselves to seek better hunting grounds, they set off at random.

Staying up all night with lions can be a real test of willpower, sitting quietly in the darkness anticipating when they will get up. Trying to stay awake with nothing to fix your mind on, nothing to distract you from whatever might be troubling you. Magnifying whatever bothers you into a major source of anxiety.

During our first few years with the lions, Anne and I intentionally restricted ourselves to studying behavior that we could observe during the day. At dawn and dusk, we watched lions mate, nurse their cubs, and feed at any kills not consumed the night before. We had equipped the back of the Land Rover with a large mattress and collapsed into a deep sleep at about eight thirty every evening, ready to start work again at dawn the next day.

We were somewhat nervous about poachers in those days. Poaching for ivory and rhino horn had suddenly increased and several animals were killed in our study area. But everyone else in the park knew we studied lions, which meant that anyone on foot who saw us parked in the middle of nowhere would assume we were parked next to something dangerous and would give us a wide berth. This knowledge gave us a sense of security, but it also led to one of the more frightening moments in my life.

One morning we found a pair of mating lions about three miles from the Seronera Lodge. A pair of nomadic males were only about a half mile away. We thought the strangers might attack the mating male and claim the female, an event that shouldn't be missed. But Anne had other things to do at home that night, and I went back alone to the mating pair. Twenty yards away from the lions, I sat in the warm evening with the car door half open, periodically stretching my legs outside to stay cool. The moon was full, the night was still, and I could hear the slightest rustle of the trees.

All of a sudden, a loud maniacal cackle rang out from just a few feet behind the car. I was convinced that the guttural, panting, snuffling sound came from a raving lunatic gone berserk; a madman out roaming the countryside, utterly oblivious to any danger from the lions. I froze for five interminable seconds. Finally, I summoned up the courage to look out the window and saw that the two lions had not even moved. Surely they had heard this horrible noise, or was it just my imagination?

The noise repeated itself, but this time in front of the car, next to the lions. They sat up and yawned as a large porcupine proceeded to erect its quills, stand up on its forepaws, and hop up and down with its hind legs, shaking its tail and emitting that terrifying sound. The lions were absolutely unimpressed, and didn't move until the porcupine came up and prodded them with its quills.

The lions had been napping on the porcupine's favorite pathway, and it had places to go, appointments to keep. Having restored order on Main Street, the porcupine continued calmly on its way and disappeared into the night. The lions went back to sleep, and my heartbeat finally returned to normal.

Oddly enough, a porcupine also gave Anne one of her worst moments in the Serengeti. Porcupines are regular nocturnal visitors to the Lion House, where they feed on our vegetable scraps, munching around outside as we eat our own dinner. One night after we had gone to bed, I heard a commotion in the living room and got up to see what was going on. Walking in with a large flashlight, I saw that one of the porcupines had somehow come in through the back door. In a demented hoarse voice, I shouted, "What are you doing here?"

The porcupine retreated through the kitchen as Anne appeared from the bedroom, white as a ghost. She couldn't imagine who I'd been talking to in such a tone of voice, but she was convinced that it must be some villain bent on destroying her family.

We have lived here long enough to know that animals are generally too preoccupied with catching the same familiar food to be interested in tasting us, and that they, too, reserve their best and their worst behavior for their own kind. Even here, in the middle of savage paradise, we fear our own species more than we fear wild animals.

It is nearly ten o'clock. I could sit here all night telling stories, but there is not much chance that these three lions are suddenly going to move their bowels. We're better off coming back first thing tomorrow morning and tracking MKP's collar. Besides, we need to go back home and do something with that fish.

TUESDAY, 5 NOVEMBER

After stumbling through breakfast, we drive off in the grey gloom of dawn. At first, the sun is only a dull orange disk on the horizon, but it quickly catches fire as it climbs up out of the thin morning haze. The morning is still cold enough for a sweater, but the blazing light soon warms the inside of the car.

Back at the Seronera River, we pick up MKP's signal and track him to a brushy part of the stream, very close to where we darted Maynard on Sunday. MKP is out of sight, but we can see three or four other lions wandering off the plain from the opposite side of the river. Doubling back from the nearest crossing, we find an adult female halfheartedly chasing away five or six hyenas that had been trying to dart in for the remains of a wildebeest kill. The carcass has been reduced to a skeleton, and the hyenas want to get at it with their bone-crushing jaws.

The female makes one last lunge at the hyenas, then walks off toward the river to rejoin the rest of her pride. The hyenas move in, tear apart the wildebeest skeleton, and scatter in all directions—loping brown figures with red thighbones, ribs, and vertebrae dangling from their twisted mouths. The female lion takes a long drink from the stream, then leaps across the narrow waterway and strolls over to the rest of her family who are already laid out under a few small acacia bushes.

Within a few yards of the pride we notice that MKP has joined the group, and we count ten lions in all: three adult females and the three young males from last night plus four young females. They are supremely uninterested in us, and we take well over an hour to identify them all. I go through my entire repertoire of insults, bleats, whistles, and groans to attract the lions' attention; we can only inspect their faces if they look up. My wildebeest calf call is reasonably effective for the first few minutes; three or four hopeful faces pop up out of the dense grass, looking expectantly beyond the car. Pam, Sarah, and Christine are already adept at drawing spots and scars, and we update ear notches on the ID cards. With this many mouths to feed at each kill, ears are frequently torn by paws laying claim to a small patch of carcass.

Each lion must have eaten about twenty pounds of wildebeest last night. Their bellies are all bulging at the seams, and every now and then someone shifts from side to side in apparent discomfort. No one had felt nature's call during the walk down to the river, but surely they will need to produce this evening. MKP and MKQ are both here, so we may hit the jackpot at last.

On the way home we find the chief park warden, William Sumai, in his office, and everyone is finally introduced. Sumai is very gracious and hospitable, but he runs the largest operation in all of Tanzania's national parks, and our activities obviously have lower priority than organizing provisions for his staff, keeping tourists happy, and catching poachers.

Sumai faces some of the most pressing problems in the entire country. Two million people, mostly poor peasants, live within fifty miles of the western park boundary. They demand animal protein, and the Serengeti is their meat shop. Pointing to the large yellow map on his office wall, Sumai describes the most recent antipoaching operations. Seronera is almost at the exact center of the park, and the most serious poaching is well to the north and west. Most of the lions we study live safely to the south and east, but the pressures from the west weigh heavily on the future of the ecosystem.

If you want to see poaching in action, just drive about twenty or thirty miles west from Seronera and walk along any riverbank. It doesn't take long to find a snare. Travel further, and you can find whole lines of snares placed strategically across each animal path. The far west of the park is riddled with hundreds of vertical pits, five feet deep, on top of which poachers place mats of dead grass and trap migrating wildebeest. Many more animals die than can be processed by the local hunters.

Several years ago, I went on patrol with a team of park rangers in the western Serengeti. After driving about five miles from their ranger post, they started searching on foot, fanning out in a broad phalanx, and walking parallel to the river bed. When they spotted the poachers, they chased them with complete abandon. The poachers were unarmed and fled into dense thickets, where they were captured and tied together. They were jailed in a nearby village the following day.

The poachers' camps were filled with long slabs of dried meat and tall stacks of bones. Poachers haul out large quantities of meat on bicycles, following animal paths through the bush. The meat is then distributed over the rest of the countryside. These are not traditional hunters, preserving traditional ways. They are businessmen. If they are convicted, they are fined. The size of the fine has not been raised for decades, and the heaviest fine is now lower than the price of a single wildebeest carcass. Getting caught is just part of the overhead.

The Tanzanian government has recently established an independent, quasi-military unit to help combat the lawlessness pervading the meat-poaching communities outside the park. Several platoons rolled into the Serengeti region last year, and meat poaching seems to be down so far this year.

Sumai is also encouraged by the fact that elephant poaching has almost ceased since the international ban on the ivory trade two years ago. Trophy poachers are in a different league from the meat poachers. Armed with semiautomatic rifles, they shoot rangers and rhinos as readily as elephants. Although their absence is good news, it comes a little too late. Trophy poaching started in the Serengeti in 1978. Since then the rhinos have been virtually exterminated and the elephant population has been reduced by ninety percent.

Most of this slaughter coincided with the socialist years during the '70s and '80s, when the Serengeti rangers were supposed to protect an area the size of Delaware with one functional vehicle. Rangers went unpaid for months on end, and fuel was scarce or simply unavailable. International agencies were either unwilling or unable to involve themselves in anti-Western Tanzania.

The Frankfurt Zoological Society was one of the few international organizations to maintain vigorous support of the Serengeti throughout the bad times as well as the good. Its founder, Bernhard Grzimek, was instrumental in establishing the present-day boundaries of the Serengeti National Park and made the place famous with his book, *Serengeti Shall Not Die*. Grzimek's eldest son died in a plane crash during an aerial survey of the park, but Grzimek continued to dedicate his life to saving the Serengeti. He appeared frequently on German television, providing his own narration for other people's wildlife films. At the end of each program, he came on screen and asked all good Germans to pull out their checkbooks and make a donation to the Serengeti. And many did.

Frankfurt Zoo did its best to help people like David Babu and William Sumai cope with the constraints imposed on them by the politicians. Now that the country is on the upswing, other organizations are becoming involved once more and the Tanzanian government is showing a renewed commitment to preserving its wildlife. Most international conservation agencies encourage tactics that could eventually replace direct military confrontation with the surrounding populace. Neighboring villagers are being taught alternative practices for utilizing wildlife. There may be short-term advantages to killing everything and eating it, but then what will your grandchildren do? What will be left? Perhaps the greatest challenge is to convince these people that they have any kind of a future at all when life is measured only one day at a time.

Back at SRI, I drape myself across the floor of the Land Rover while installing a two-way radio. A whiff of something keeps wafting up from somewhere near the front wheels. It smells as if we had run over a rotten carcass, but the tires are clean. Has a hyrax fallen into the rainbarrel at the corner of the house? As I search around between car and water tank, Christine rushes over from next door and breathlessly announces that Erin has finally dropped her bomb: she has decided to end her participation in the lion project, and she wants to talk to us immediately.

Erin, Christine, and I sit down together in the living room of the Lion House. In a tense, angry voice, Erin lets it all out. She resents my attitude toward her. I treat her like a second-class citizen; I don't talk to her, I don't listen to her. Christine is too German, too formal, and doesn't take her seriously. Christine is from Oxford. When Erin went to Oxford with Rob for a month, everyone looked down on her because she was Australian. No one would listen to her. We have all treated her so badly that she has hidden her forty fecal samples in another house at SRI. She may try to do something with them when she gets back to Australia, or she may just dispose of them.

My turn. I may not have spent much time with Erin, but I had expected her to find a niche within the parasitology group at Oxford. I was only there on sabbatical myself, a visitor. I had arranged for her to have her own office when she arrived. If I seemed distant, it was hard to be otherwise when she was always so negative.

Christine's turn. She had spent a lot of time with Erin and had arranged various opportunities for Erin to visit other parasitologists around Britain. Lots of people had argued with Erin during her stay at Oxford, but Erin had said lots of provocative things; graduate students at Oxford tend to have strong opinions. Erin is too sensitive about being Australian—Oxford is full of successful Australians.

I had encouraged Erin to participate in the lion project because researchers' spouses often become utterly bored in the middle of paradise. Erin had developed an interest in parasitology in Australia. She has an active mind and wanted something to do. After ten months in the Serengeti, though, she feels that any moron can follow a lion around and pick up its shit.

The loss of forty samples would be a major blow to Christine's project and we don't want Erin to flush them. Outside, I find Rob and ask if Erin should just go back to Australia now instead of waiting until January. The isolation here seems to have affected her perceptions, a phenomenon known in East Africa as "being bushed." There are so few people to talk to in a place like this; you are constantly assailed by random memories of conversations both ancient and recent. Statements lose their context, compliments become criticisms, offhand remarks become solemn proclamations. The same old arguments run round and round in your head. There is nothing to bring you out of yourself.

Rob is in an awkward position and doesn't want to be caught in the line of fire. He counsels patience and says that Erin has just gotten worked up by bringing it all out into the open. It is very important that we listen to what she has to say. She has thought hard about the parasite studies and wants to feel that she had made an intellectual contribution to the overall project.

Taking a deep breath, I go back inside the Lion House and sit down again. I apologize very carefully to Erin for any slight that I may have committed. "If I did seem offhand and distant, I am very sorry. Your participation is certainly highly valued, and we are very keen to hear your suggestions for where we go from here. You know more about collecting these samples than anyone else; we need your help." All true, all sincerely meant.

Then Erin launches into a speech she has obviously rehearsed in her mind for days, if not weeks. Certain prides "bog" more often than others; the lions usually bog just after sunset or first thing in the morning. "They need to walk a short way to get the old bowels in motion, so look out for a squat after they start traveling. Watch the prides on the plains when there are tons of wildebeest out there; the more food in, the more crap out. When the wildebeest are up in the north, concentrate on one of the prides around SRI because they eat so much buffalo. Big prides are better than small prides; the more lions there are, the more likely that someone will need to bog."

All very positive and well-intentioned. But then the finale, laced with venom: "Now, Christine, graduate students should collect all their own specimens in the field and then do their own analysis. You should stay out here for the next two years to gather your own data, then go back to Oxford."

Calmly avoiding any hint of sarcasm or condescension, I thank Erin for her suggestions and ideas.

"However, I am the one who decides how long anyone can stay out here, and we do not have space for Christine to stay any longer. Her work will be enormously time-consuming once she goes back to Oxford. She will have to complete her thesis within three years, and most other graduate students rely on assistants these days. We are grateful for the rest of your suggestions, and we will consult with you each day."

End of session. Pam and Sarah have driven over, and we have to rush back out to the lions. Christine and I leave Erin on better terms than before, and she seems somewhat mollified. She has decided to continue to help collect additional samples for the duration of our stay ("What else is there to do out here?"), but she has not yet decided to return the original forty samples. We are made to understand that they are being held for ransom.

Sitting in the driver's seat of the Land Rover, I feel utterly exhausted for the first time in the past ten days. But we have to keep up the pace, have to get back to the lions before they bog. "Don't ever drive like this," I shout over my shoulder as we return to the Seronera River in much less time than I care to admit.

We arrive just in time to see a young female lion stalking something on the other side of the river. She creeps along for a few yards with her head held low, then freezes for several long, tense seconds. The rest of her pride are all lying flat out asleep in the open, and as she moves slowly forward into a stand of tall grass, a reedbuck suddenly bursts out onto the open plain, stops, turns, and whistles a piercing alarm call at the lions. The young female stands erect, looks around for a minute, then comes back to the rest of the pride and conks out again.

We have all been sitting perfectly still during this tiny life-or-death drama, and it has certainly helped to throw the day into perspective. Conflict of interest is the rule, not the exception. Everyone has his or her own motives and desires. After all, Erin has her reasons, too.

On one level we are working for a common goal, but on another there is nothing but dispute and dissatisfaction over the most insignificant issues. Life in the bush can bring out the worst in the rugged individual who wants to do everything for himself, the Lone Ranger riding over the hill in a cloud of dust. But such people don't often seem to realize the extent to which their own success ultimately depends on the goodwill of the very persons they scorn. They seem oblivious of their own mutual dependencies.

People often find it hard enough to get along together back in civilization, but we can at least keep our professional and personal lives separate back there. Out here, the professional life *is* the personal. It is unavoidable. We are thrown together with a set of individuals that we would otherwise never invite into our homes. Here they end up living in our home.

It is now quite dark. Distant features have blended into the dusk; the foreground has turned a dull, blank grey. I take the nightscope out of its case, point it in the approximate direction of the lions, and then gaze into the phosphorescent screen, which glows bright green inside the car. That vague blob is a foot jutting up from the matted grass, the rest of the lion's body is hidden somewhere beneath a fuzzy tangle of thorny branches. But without shadows there is little sense of depth. Everyone eagerly takes a turn looking through the large eyepiece but is soon disappointed by the blurry, flat image.

Occasional breaths of wind carry the same smell of death that eluded me in the afternoon. There must be something under the car somewhere. We drive over to an open spot about fifty yards away, tracking MKP, and find him lying in a heap with three or four other lions. The nightscope works best on a moonlit night, but the moon is new this evening and has already set. We back the car around slightly to watch the lions with our parking lights.

Gunning the engine sends a strong whiff of dead animal into the car. MKP is attracted by the smell and he gets up to come sniff at us. He walks all the way around, sniffing, looking occasionally through the windows. My passengers sit in the darkness, frozen by a mixture of terror and elation. The lion strides toward the front of the car and starts to chew on the tire. I reach outside and swat him on the butt. He jumps away half a step, regains his composure, then calmly walks back to his buddies and flops out again.

MKP has made it official: we have something dead under the bonnet of the car. Turning on the engine creates a gust of wind from the engine fan—and then I remember the squealing noise as we were leaving the house Sunday evening, when the alternator seemed to have failed. A hyrax must have been warming itself on the engine near the fan belt. We have driven far enough to have almost cooked it, but its innards must be well-rotted by now.

I turn the car ninety degrees and blow another massive whiff of rotten animal into the car. Judging by the sleepiness of the lions, we won't see any action here in the next few hours. Bowels will not be emptied until morning. I am absolutely wiped out by the confrontation with Erin and declare it time to head home before the whole pride decides to come gnaw on the car.

Christine uses a spotlight to look for night creatures while we drive home. Sometimes we can see the large, shining eyes of bush babies in the acacia trees beside the rivers, or nocturnal mongooses out foraging in the grass. Tonight we only spot an owl, perched on an acacia branch above the road, feeding on a small rodent. It blinks its wide eyes at us a few times, then resumes its feast. Another minor conflict has been resolved.

As soon as we arrive home, I open the bonnet and, oh Christ, I was right. A hyrax in unspeakable condition, but at least in one piece. I grab the corpse, yank it out, and fling it into the bushes. Thank heaven I'm surrounded by vegetarians.

WEDNESDAY, 6 NOVEMBER

This morning Pam is behind the wheel for the first time, and she drives us toward a final showdown with the heart-eaters. Along the Seronera River, she hears a faint beep very close to where we left the lions last night. We circle, laboriously at first, as she discovers the difficulties of radio-tracking and driving at the same time.

"Now remember where to cross the river. Mind the ruts in the road . . . Go ahead and leave the track, but look out for pig holes . . . Drive over there where the grass is shortest . . . Is the signal getting any stronger? Try circling again . . . Got a new direction? There they are! Well done!"

At least a dozen lions are lying in a tight bundle in the middle of the plain, not far from the Masai kopjes. Most of the adult females are here and so are all of the subadults. John, Maynard, and Smith are tending to business somewhere else today.

Now for the hard part. Identify every single one of them. Make all the noises to wake them up. Drive around to view each face. Try to remember who is where. Aside from the mild impatience to finish off the task, I am happy just to sit back and listen to Pam and Sarah figure it all out.

One of the subadult females, MKS, appears to be ill. She seems uncertain on her feet as she walks about fifteen yards away from the others. She is very pale, her fur almost white. Coat color doesn't fade with illness, but her paleness makes her look frail. She stands still for a moment, swaying slightly, then sits down and looks around as if she were dizzy.

Several of the other subadults walk a few paces at random before settling down on more desirable pieces of real estate. One of the subadults stops her meandering, squats, and produces an almighty turd.

Loud cheers from the backseat of the car. But hold on, who is she? A young female, MKN. Then the sick female, MKS, gets up and staggers toward the group. She stops, too, and with the same look of quiet concentration, also proceeds to relieve herself. She's got diarrhea! Maybe that frothy black pudding holds the key to her illness.

Don't forget where each prize is located—one is by a small tussock of grass, the other is in a patch of bare soil. Yikes! There goes a young male. Nuts! It's MKR, the only male that hadn't eaten an ox heart. But now it's like a chain reaction—MKQ is doing his bit for science. The number one heart-eater, he's our top priority.

It is getting hard to remember who did what where. We drive past the sick female's specimen and drop, appropriately enough, a small streamer of light blue toilet paper to mark the spot.

By the time we arrive at MKR's sample, several subadults have gone over to play with the toilet paper. One of them picks it up and the rest quickly give chase. Half a dozen young lions start romping around, cutting back and forth, darting in at the trailing streamer, until it is ripped to pieces and everyone loses interest. The whole group starts walking down toward the river, heading for shade. Various subadults try unsuccessfully to get their mothers to play tag. About a hundred yards from the car, MKP makes his contribution, and we look desperately for some sort of landmark.

After the lions are a safe distance away, the scat squad is out of the car with wooden tongue depressors and plastic pots. Everyone is wearing gloves. Lion shit is full of liver flukes, hookworm, giardia, and the like. But the most hazardous pathogens are echinococcus, a tapeworm that causes cysts the size of a basketball, and toxoplasmosis, which can be extremely damaging to small children and AIDS victims.

Christine scoops up a sample from each bog, and Sarah labels each pot with the name of the author and the date of the production. Pam keeps a bearing on the various small piles that are dotted around on the plain. This seems like a good time for me to get out of the car, stroll upwind, and admire the view.

Thirty yards from the car the air is fresh and the view is fine. The Masai kopjes crown the ridge top to the right. Across the valley to the left are the grey gothic towers of Prinz's kopjes and the converging tributaries of the upper Wandamu River. A troop of baboons is down by the stream along with several small herds of waterbuck and topi. The back-lit grass glows only slightly green. The wind is starting to blow, the sky cloudless, the ground dry. Another rainless day is in store.

Like farmers, we think a lot about the weather out here. It is the rain that drives the migration, all those thundering herds charging to wherever the grass is greenest. Some years, the rain falls in just the right pattern to keep the great migratory herds around Seronera the whole year round. Those have always been bumper years for lion cubs. Nineteen sixty-six was a good vintage, so were 1973 and 1984. Meat was plentiful and no one starved.

There are two rainy seasons in the Serengeti each year, the "short rains" from November to Christmas, and the "long rains" of March through May. All the browsing and grazing mammals look pretty sorry for themselves by the end of the long dry season. By late October the gazelle are usually mangy and the buffalo are bony. The wildebeest start to drop like flies. The smaller birds are pretty tatty, too, waiting for the rains before molting into their breeding plumage.

The rain literally switches the plants on and provokes an explosive burst of life—lilies spring forth from the bare soil, trees flower, the ground comes alive with bugs. At night, electric lights attract enormous swarms of insects: flying ants, termites, dung beetles, burying beetles. Beetles, beetles, and ever more beetles emerge from the wet earth, slipping in through the smallest cracks of a house or car, burrowing in our hair, tunneling under our clothes, marching on the dinner plates—we call it beetlemania. Either we go to bed early or we just sit in the dark and listen to the scraping of all those tiny legs.

The short rains are erratic and sometimes fail altogether. Last week's showers were only a flash in the pan. Over the past few days, the combination of fierce sunshine, harsh wind, and high altitude have already sucked the moisture from the soil. Most of the trees have sent out new leaves, and we can still see an occasional lily, bright pink in the grass, but everything else is on hold, waiting for a proper deluge. Enthusiasm is starting to wane; it is a reversible spring.

It seems unlikely that the rains will be sufficient to pull the migration back to the plains while I'm still out here this year. Although I would be disappointed to miss the spectacle, it will be easier to get around if the rains do fail—no getting stuck in the mud, no raging rivers. I can look forward to a few weeks of comfort and calm instead of beauty and mayhem. After all, I'm supposed to have come out here to train these people, not to indulge in the cheap thrills of living in Africa.

The lions, already a good half mile away, have almost arrived at the trees by the river. In spite of all that open expanse, they are bunched together like fish in a shoal, a yellow school plodding along the surface of a fading green sea. Lions are social animals all right, but from up here their togetherness looks like a protective formation, each one staying close to the others, anticipating an attack by some unseen enemy. United we stand.

These subadults were all born within a few months of each other. Their mothers brought them together when they were a few weeks old, and they have spent virtually their entire lives as part of a nursery group. They have survived great odds to make it this far, but there are many more hazards in store.

A female lion becomes secretive during her last weeks of pregnancy, and then gives birth in the dense vegetation of a kopje or riverbed. Each female has a preferred den site, often the spot where she herself was born. After giving birth, the mother spends most of the next few weeks alone with her cubs, but leaves them periodically to go hunting. She occasionally moves her cubs from one den to another—the familiar image of the lioness carrying her cub by the scruff of its neck—and the cubs stay hidden until they are four to six weeks old. Small cubs alone in their den make little effort to escape. Their only defense is to remain undetected.

As soon as her cubs are old enough to move around on their own, the mother brings them out to join all of the other young cubs in the pride. Births are often synchronous within the same pride, so most cubs end up being reared in a "crèche" formed by two to four females. Mothers nurse until their cubs are about six to eight months old, and the crèche persists until the mothers are ready to breed again, when the cubs are about a year and a half old.

The mothers of the crèche are constant companions and serve as the social focus of the pride. The popular conception of the large pride of lions complete with mothers, males, and masses of cubs is based on the occasional pride with a large crèche. Childless females are much more flexible in their groupings and frequently travel alone; they often use a crèche as a meeting point to rejoin other pridemates.

Anne and I originally had come out to study the lions because we were so intrigued by their communal cub rearing. Lions are famous for nursing each other's young. A number of insects and birds are communal breeders, laying their eggs together and looking after the collective brood, but such behavior is quite rare in mammals.

Is communal nursing truly cooperative in lions? Are mothers actually assisting each other with the constant chore of child care? Or are they together for some other reason? Communal nursing might even be parasitic; infants of many species are known to steal milk from females that are not their mothers. Can female lions even recognize their own young?

Our first few years in the Serengeti were fairly hard ones for the lions, and we were often frustrated by a shortage of surviving cubs. Insufficient rainfall kept the migratory herds away from the Seronera region for much of the year, and many small cubs starved. There was also a great deal of social disruption due to a recent increase in the lion population, which meant that many more cubs were lost. We checked up on each pride every few days, looking for females that had recently given birth; the suckling of the cubs leaves conspicuous brown stains in the fur around the mother's teats. Time and again nursing mothers lost their cubs before they could be brought out to join the rest of the pride.

We were able to study only two crèches during our first year in the Serengeti, but conditions improved in subsequent years, and eventually we were able to study a dozen communal litters as well as the litters of several mothers (singletons) that attempted to rear cubs on their own. We watched nursing behavior for well over a thousand hours, sitting parked in our car only a few yards from the crèche while mothers snoozed and cubs tried their luck at obtaining milk from each female.

This project was a major focus of Anne's research, and I often acted as secretary and scorekeeper, filling in data sheets while she commented on the complex interactions between mothers and cubs from one moment to the next.

Anne: "Here comes B to nurse from Lola, and A has just gotten off Lolly."

Me: "Okay, that leaves C and D still on Lolly, and E and F must still be playing with Laini."

"Yes. Now D is off Lolly, and Laini is snarling at him as he walks by."

"What's Loliona up to?"

"She's lying on her front; no one has tried her for a while."

It didn't take long to discover that female lions can indeed recognize their young. They are much more tolerant of their own cubs' attempts to suckle than those of other cubs. If another female's cub tries to suckle from her, a female will often snarl and lunge at it; all those god-awful teeth swooping down to keep the little pest off. A mother also licks her own cub much more than she licks other cubs. Her powerful tongue can virtually lift the cub off the ground, cleaning its fur and removing small ticks and mites.

So communal nursing is not the result of mere stupidity. Nevertheless, almost thirty percent of a mother's milk goes to cubs that are not her own. Anne looked at our data from many different angles, trying to sort out which of a dozen or so factors were most important in influencing the extent to which females nurse each other's offspring. As it turned out, the most striking aspect of milk distribution within each crèche is that the females who are most generous with their milk are the ones with the least to lose.

Cubs often manage to nurse from the "wrong" mother when the female is not really paying any attention to what is going on. After all, mom has been out hunting all night and she needs to nap as well as feed the kids. Females that have pooled their cubs with their closest female relatives show less concern about the losses—at least the milk is going to a close genetic relative—and a familial crèche is a very laid-back group. On the other hand, if the partners are only distant cousins, mothers are more alert and nurse mostly their own cubs.

If a female has only one or two cubs, she can (and does) spare more of her milk than a female that is trying to raise a litter of three or four. Females produce roughly the same amount of milk regardless of the number of cubs in their own litter, so a mother with a small litter has more milk to spare. Female lions have four teats, and an "only cub" is often very popular at mealtime, surrounded by friends at the dinner table.

So why do mothers allow other cubs to nurse from them? Why put themselves in a situation where they can be parasitized? Even if generosity reflects kinship or small litter size, a female could direct *all* of her milk to her own offspring by simply rearing them apart from the rest of the pride. That's what females in other feline species do. What do female lions gain from forming a crèche? Do mothers in crèches acquire more food for their young? Perhaps they can afford to be relatively indiscriminate with their milk if groups are able to hunt more effectively and catch more prey.

We know that rates of food intake influence the quantity of milk a lion produces. We have milked several lactating females, and mothers that had recently eaten larger meals produced more milk. However, by various estimates of the amount of meat that a female lion eats in a day, it appears that mothers in crèches generally take in less food than mothers that forage alone. Thus cubs of communal litters probably receive less milk than the cubs of singletons.

Perhaps the mothers tolerate nursing by the wrong cubs because they have a variable milk supply? Maybe one female feeds the whole batch when she has had a large meal so that the cubs can have a steady diet when their own mother comes home empty? Or perhaps the mothers pool their cubs so that each mother can take turns to go off hunting?

But communal mothers remain together even when they leave their cubs: they hunt together, feed together, and return to their cubs together. All the mothers feed from the same carcasses and receive comparable portions of the meat. When one mother goes hungry, they all do. Thus the food supply is no more steady for a cub in a crèche than for a cub reared only by its own mother. Nor do mothers leave a baby-sitter with the crèche.

Anne's work shows that communal nursing is the price mothers must pay for being together so much. The cubs often suffer nutritional costs from being in a crèche, but mothers are by far the most gregarious females in the pride. Females who don't have cubs are much less sociable. Why do mothers stay together so much? What other advantages do they gain from forming a crèche that could overcome the nutritional costs to their cubs?

Communal rearing does indeed confer a very real advantage, but the benefit comes from the fact that larger groups of females are better able to protect their cubs from a constant danger. That danger does not come from some other species but from other lions. The motivating force is sex, and the ultimate targets are the mothers themselves.

A mother must devote all of her efforts to her current batch of young. She weans her cubs when they are about six months old but does not resume breeding until her current brood become reasonably competent at hunting and fending for themselves. However, should her cubs die at any point before their second birthday, the mother will start mating within days. A lioness wastes no time on mourning. She has to start over and breed again. But if her cubs survive, they will have served as unwitting contraceptives.

Although mothers spend nearly two years rearing each batch of cubs, male lions are in a hurry: they are under a lot of pressure. Male coalitions compete intensely with each other for access to a pride, but competition does not end once a coalition gains residence. Other males are always out there looking for a chance to breed, and the resident males have to be prepared to defend their mating rights at all times. Roaring and chasing away strangers is what a breeding male's life is all about.

All resident male coalitions are eventually defeated by younger and more vigorous challengers. No one stays on top forever. But the offspring of the displaced coalition are merely an impediment to the breeding success of the usurping males. The new males have to start their own breeding effort as soon as possible before they too are ousted. They want the females to resume mating immediately, but mothers with dependent young are unable to conceive. To father cubs of their own, the usurping males must therefore kill the offspring of the previous coalition.

Killing cubs less than four months old speeds up the mothers' reproduction by about eight months. Eight months is a very long time for a male lion. Over his entire adult life, he may reside in a pride for no more than two years. He has no time to waste being a stepfather.

Infanticide in lions has been observed only about a dozen times, but small cubs invariably disappear whenever a new coalition takes over a pride. We know of hundreds of cubs that have died in this way; infanticide appears to be the normal course of events.

In 1987 filmmakers Richard Matthews and Samantha Purdy started filming a television documentary about our lion research. During the planning stages, they asked what we considered to be the highest priority for the film. Without hesitation, we told them to try to film infanticide. Infanticide might make unsettling television, but it is essential to understanding many of the most fascinating aspects of lion life.

We warned them, though, that infanticide could happen in a second, and that it usually occurred at night. Direct observations were rare, and we usually inferred infanticide from the disappearance of small cubs. Even if Richard and Samantha did manage to see it, it might be impossible to film; they might waste months in the effort.

They decided to take the risk, and we suggested they concentrate on the Kibumbu and Boma prides. The resident males in these two prides had recently disappeared, and each pride contained a lone mother with small, vulnerable cubs. We were certain that nomadic males would soon move in, eager to claim the abandoned females. Each day Richard and Samantha headed out before dawn, checking up on both prides.

After a few weeks, a coalition of nine young males began to zero in on the Kibumbu pride, and several females started to mate with them. The Kibumbu mother became harder to find, withdrawing from the rest of her pride and staying a safe distance from the males. Late one afternoon, Richard and Samantha arrived to find one of the new males sitting about twenty-five yards from the singleton mother and her two cubs.

The mother snarled viciously at the male whenever he moved. She stood straddling her cubs, trying to keep them hidden in the tall grass. The male showed no impatience, no hostility, and stayed put until well after dark.

Late that night, the male finally got up and charged straight at the mother and cubs. She had no choice but to run; she was less than two-thirds his size. The male's head disappeared into the grass next to the retreating female. He pulled up one mangled cub, then the other. Two days later, the mother came into estrus and mated with the murderer.

Though visibly disturbed by the slaughter, Richard and Samantha were frustrated by the poor visibility and the bad light that had made it impossible to film the attack. They turned their attention to the lone mother in the Boma pride. This was their second chance, but it might also be their last.

Richard and Samantha watched her every day for three weeks, but nothing happened. The cubs were growing well, the mother was finding plenty to eat, and there were no males in the area. We began to wonder if we should change the direction of the film.

Finally, Richard and Samantha arrived one morning to see the Boma mother slowly retreating from an advancing group of lions. Behind three invading females came Smith, of John, Maynard, and Smith. The mother had been reluctantly avoiding the trio of females, but once she saw Smith, she ran off at great speed, leaving her three young cubs behind. The females calmly walked toward the cubs, looked at them briefly, then went after the mother, intent on their territorial dispute. When the cubs saw Smith, they became agitated for a few seconds, then crouched down and froze.

Smith walked to within forty yards of the cubs, stopped, sat, and watched the females trotting off into the distance. After about half an hour, the wind suddenly shifted. Smith detected the cubs' scent. He began to stalk them, slowly, in a half crouch. Then he growled and charged straight at them. The cubs scattered, but they had no chance. Smith picked up the first cub by the stomach, bit hard, and shook it. He seized the second cub by the head and crushed its skull. The third cub had managed to move a few yards away, but Smith quickly found and killed it, too. All three cubs lay dead. Smith lifted one body and carried it to the shade of a nearby acacia tree. He ate all but the head.

We felt a strong mixture of elation and revulsion at Richard and Samantha's success. When the film commentary was eventually recorded by the actress Lindsay Wagner, her voice cracked with emotion during the infanticide sequence. She wasn't acting. Lion cubs are so endearing that when a huge male comes up and crunches one, it is impossible to remain unmoved.

But all male lions do this. Every breeding male in the Serengeti has probably killed at least one cub in his life. Every lion in the world has a father who is a murderer. A good citizen in this society would leave very few offspring; virtue is weeded out.

Viewing the film ourselves, Anne and I were struck by the way Smith growled during his attacks on the cubs. Lions always remain silent while attacking their prey. This was an act of aggression. Smith was also careful to kill all three cubs—the only way to accelerate the female's reproduction is to kill off her entire litter—but he didn't eat them all. The ultimate purpose of infanticide is sex, not food.

Richard and Samantha's observations emphasize the difficulties single mothers face in defending their cubs against infanticidal males. After all, males weigh fifty percent more than females, and the cubs are an easy target. By contrast, groups of females are much more effective at defending their young. By banding together, the females can overcome the difference in size and help at least some of their cubs survive.

During our early days in the Serengeti, Anne and I observed a crèche in a small pride that contained only three young females, their six cubs, and a pair of elderly males. We first found them lounging serenely under a large sausage tree by the Seronera River. Prey was plentiful, everyone was fat and healthy, the family was complete. Only a few miles upstream, however, a large neighboring pride had recently been taken over by a coalition of four males. We arrived early one morning to find the quartet chasing out the resident pair from the small pride. The invaders roared as they trotted along, running for over a mile, foaming at the mouth. The three mothers of the small pride retreated in formation around their cubs and took them to the thick vegetation by the river.

As the invading male quartet returned to claim its prize, two of the mothers ran out snarling. The third mother moved along in the bushes, leading the cubs to safety. After an intense clash with the advancing females, two of the males went after the third female. By the time they arrived at the river, she had already hidden the cubs somewhere else and had come out to confront the males. Three of the males stayed in a close circle around her for the rest of the day, holding her prisoner.

Late that night the four males went home to their own pride and the mothers cautiously reunited. Amazingly, all six cubs had survived the encounter. The two fathers managed to return a few weeks later, and the pride eventually reared three of the cubs. But the lions never regained the calm composure we saw that first day. The mothers were jumpy and moved from one part of their range to another, never staying in any one place for long. We rechristened the pride "Shambles."

The advantages that mothers gain from defense and distraction explain why the crèche is so militantly gregarious: the *females* are the ultimate goal of the males' attacks. If one mother stayed behind with the cubs while the rest went hunting, she could easily be overpowered. By staying together, the females gain maximum protection. They only resume their separate lives when their cubs are independent and they themselves are ready to start mating once more.

The youngsters of the Masai Kopje pride are almost fledged; some of the mothers are already pregnant again. John, Maynard, and Smith have held on long enough to father a second set of young. Even at this age, though, the subadults are still at risk from a male takeover. Incoming males evict all of the subadults in their new pride. Subadults compete with younger cubs for food, and whereas mothers and fathers try to balance the conflicting demands between their older and younger offspring, stepfathers want to make sure that all resources go to their own cubs. Once evicted, the young males must begin their lives as nomads; young females try to establish a new pride. If they are evicted at too early an age, they will starve.

Although a mother may have to defend her young a dozen or so times in the two years needed to raise them, we can hardly expect to go out every evening and see a slam-bang encounter. In order to gain a more complete picture of the mothers' defensive behavior, we eventually resorted to playback experiments. The recorded roar of a male from a neighboring pride elicits a dramatic response: the mothers bunch together in a tight group and sometimes charge a step or two toward the speaker, obviously ready to defend their young. Then the whole batch nervously retreats, sometimes moving several miles overnight.

After Anne and I completed a series of pilot experiments, Karen McComb performed many more playbacks of this sort, and she also played back roars of the cubs' fathers. When the mothers and cubs heard the roars of the resident males, they would just roll over and go back to sleep. Mothers were only perturbed by the roar of a male that would benefit from killing their cubs; the sound of the cubs' fathers was no cause for alarm.

Communal cub rearing has several ironic aspects. First, the mere fact that males come into a pride and clear out all of the cubs synchronizes the reproduction of the mothers. They all lose their cubs at once, their reproductive clocks are simultaneously set back to zero, and they subsequently give birth over a fairly short period. Synchronous breeding makes the mothers more likely to pool their cubs with those of other females, thus setting up the defensive formation that prevents a second successive loss from infanticide.

Second, and rather more macabre, the very formation of a large crèche provides the foundation of a very large cohort of young. The males of the cohort will stay together as a coalition after they've grown up. The larger the coalition, the better their chances to take over a new pride some day—and successfully kill someone else's cubs.

Thus the very defensive formation of the females provides the foundation of an extremely effective attack force once their own sons grow up and set out to conquer the world. This is the twist in the tale that reminds me of the endings of so many creepy horror movies: a tiny piece of the fallen monster regenerates to produce another monster in the fade-out of the film.

As I watch the lions walk across the plain today, I think about the complex relationships that they have developed with each other over the past two years. They have had their share of petty bickering over small matters such as the fair distribution of milk and the endless squabbling for meat at kills.

But at the big moment, when the villain strikes from the middle of the darkness, they all pull together. It is life or death, and there is never any doubt about what to do.

THURSDAY, 7 NOVEMBER

While we sit idling on the Seronera airstrip, Sebastian Tham finishes checking his gauges and dials and then broadcasts details of our imminent departure over the radio. "This is 5 Hotel Zulu Oscar Oscar, preparing to take off from Seronera." All private aircraft in Tanzania are listed as "5H." Sebastian works for the Frankfurt Zoological Society, organizing the Seronera garage and flying to each distant ranger post whenever an anti-poaching vehicle expires.

Tests complete, the skies empty, the engine begins to roar, and our tiny four-seater skims the surface of the ground as Sebastian chases a small herd of topi off the strip. Runway clear, we come charging back in the opposite direction. I put on a heavy pair of headphones to cut out the noise and prepare to listen for radio-collared lions.

Quickly airborne, we bank off to the right, then fly along the course of the Seronera River as it snakes and coils beneath us. Dark green marshes and palm groves stand out against the yellow grass and pink soil. A giraffe ambles slowly toward the stream, casting its long shadow like a sundial. Dozens of narrow game trails converge upon the riverbanks. There is surprisingly little vegetation away from the river. The grass seems plentiful only when viewed obliquely from ground level. The crowns of the umbrella acacias are so lacy that you can see straight through them. Their spidery shadows repeat each tiny detail of branch and twig on the ground below.

An antenna is attached to each one of our wings, connected by thin black cables to a small black box on my lap. I flick a switch back and forth, listening for beeps from each antenna: left, right, left, right, left, right. When the signal is equally strong from both antennas, we know we are flying directly toward the target.

Two miles down the river, I hear the steady beep of CSN, the Campsite female we tracked our first night. She is still several miles away, but it sounds as if she is at her den again today. The signal grows louder and louder as we head north. Sebastian brings us down to within a hundred feet of the ground. The transmitter has a greater range at higher altitudes, but the signal is less directional, so we have to fly lower as we get closer.

Directly above the river, more or less over the same spot where we found her last week, the beeps become almost deafeningly loud. As soon as we pass above her, the signals start to fade, and Sebastian punches the Global Positioning System gizmo to record her precise latitude and longitude. As we rapidly climb, I cancel her frequency from the receiver and start scanning for the next two collared lions.

Above the summit of Nyaraswiga Hill, more and more river systems come into view to the north. More groves, more flowing streams. Not far from here the grass is still green and we can see a few large herds of wildebeest and zebra, their dark bodies filing across the grasslands like ants. It would be nice to be able to enjoy the view, but we've already picked up another signal.

"She's somewhere off to the right . . . Sounds like she's over there beyond the Sangere River."

Left, RIGHT, left, RIGHT, LEFT, RIGHT, LEFT, RIGHT.

"Straight ahead now. Go lower for a better fix. Somewhere down there?"

"I'm not sure. Let's go back again and give it another pass."

"A little further to the left this time . . . Okay . . . Lower . . . Now down there somewhere."

LEFT, RIGHT.

"Yes, that got her."

"Back up again . . . Hold it . . . I've already picked up the next one . . . Yeah, over there by that kopje . . . Back down, hard right."

Your body wants to stay up in the sky, but you are being yanked down by the seat belt. Now you're being pressed hard against the window, now up, now down. The world flips on its side, then rebounds. Your neck asks for mercy. This is not your ordinary day at the office, dear.

The noise, the tension, the vaguely sweet smell of aviation fuel, and I'll admit that I feel slightly queasy. I never feel much worse than this, but it will take a few hours to recover once we have landed. Christine seems to be taking it all in stride, but Pam is fairly close to the limit. She drew lots with Sarah last night to become the third passenger. She regrets it now.

The pilot always brings along a stack of small bags, prepared for the worst. You never can tell who is going to lose it on a flight like this. One of my former graduate students, Dave Scheel, needed to go along on each flight to measure the distribution of prey animals while we tracked the lions. Dave is a sturdy character on the ground, solidly built, with a bright red beard and hair. But by the time he was fifty feet in the air, he would be white as a sheet. After an hour or so, his face would really and truly turn green.

Heading east toward Turner's Springs, we quickly find the Loliondoes. They are out in the open, and L23 is with a large batch of companions. We turn to the south and east toward the open plains. From now on, the prides will be few and far between. The plains prides survive precariously in vast tracts of grassland that attract the migrating herds of prey for only a few months each year and support only a small number of warthog and gazelle the rest of the time. Occasional feast punctuates chronic famine. Lions in the woodlands need only about twenty square miles to find enough food, but prides out here may range over two hundred square miles.

Higher and higher we go. With enough altitude, we should be able to track every lion within twenty miles. We fly east along the spindly drainage line of the Ngare Nanyuki, the river that marks the limits of the plains from Turner's Springs to the eastern boundary of the park. A few patches of green grass are still scattered around here, but the vegetation is much sparser than at Seronera.

I suddenly pick up the Kibumbu pride well north of the Ngare Nanyuki. The Kibumbus have rarely been seen in the past two years, so I want to census them from the air. After about five gut-wrenching passes, we finally see four lions in the shade of a small bush. I've never been this far beyond the Ngare Nanyuki before, so we mark the spot with the GPS and look for river crossings to drive back in the afternoon.

Finally we head out for the real hardship cases on the eastern plains. These are the outer fringe of the Serengeti lion population, the prides that somehow survive in the heart of the rain shadow. During the wet season the Sametu, Barafu, and Gol prides cling to their eponymous kopjes. During the dry months they break loose from their moorings and wander far and wide.

From two or three thousand feet in the sky, it looks as if we are on a mission to Mars. Brown and dull, the ground below is too featureless to belong to this world. The scattered grey clusters of kopjes mark the devastated landscape like an ancient warning in braille: "Earthlings, go home."

The stark slopes of the Gol mountains stand out on the eastern horizon and the Crater highlands rise farther to the south. Against the wastelands below, the massif looks like the gateway to Shangri-la. The mountains define the eastern limits of the ecosystem and make the place seem much larger than it actually is. The Serengeti is vast, but from here it seems a world unto itself.

Last week's arrival of the migratory herds has pulled all the lions north: the Sametus are at the northern edge of their range, the Barafus have moved all the way up to a marsh along the Ngare Nanyuki, and the Gols have shifted up into the Sametus' usual range. In this open terrain, we need only a single dive to spot each pride, but we start from a much greater height and Sebastian corrects course on the way. While hurtling down like a crippled jetliner we have plenty of time to contemplate the afterlife. Obliteration would at least bring sudden relief. We easily spot each pride, and Sebastian uses our accumulated speed to launch right back up into the heavens, kamikaze over and over again.

Maintaining a moderate height, we pass over the Gol kopjes, bleak islands in an endless, yellow sea. Among each great pile of rocks, leafless grey trees stand out like tattered shipwrecks on the Skeleton Coast. We turn west and fly over absolutely nothing until the dark mound of Naabi Hill, where we turn back north again to head toward the Simba kopjes. Down on the Arusha road, huge plumes of dust mark the passage of this week's tourists, heading for town in their minibuses.

At last we come into range of the Simba pride and our survey is almost complete. It is green again over here, and the Simbas are feeding on a wildebeest, its guts gone, its limbs torn from the trunk. To the west, the Nyaraboro plateau stands defiantly against the prevailing east wind. Bushes dot the hills, trees line the streams, the kopjes are bursting with life: take one large pile of rocks, add water, and stand back. Within half an hour we have flown in and out of the rain shadow, all the way over to Mars and back.

The Plains pride is hidden in the tall grass close to the Loyangalani River. Lake Magadi glistens blue in the background; a small pink flock of flamingos lines the northern shore. After climbing from our final swoop, we start back toward Seronera, but I want to look for one last animal before we land, a female that split off from the Campsite pride a few years ago. I've tracked her throughout the entire flight but want to try searching from a greater height above Nyaraswiga. We climb up and up until theoretically we should be able to see Lake Victoria, but there is too much haze today, and we have no luck finding the lost female, either.

To the west of Nyaraswiga, the landscape is wild and beautiful. Rocky cliffs tumble down into idyllic pastures watered by cascading springs. To the north and east lie parallel streams and wild, unspoiled country. But to the northwest, even from here, you can see something else—the habitation that stretches along the western boundary of the park all the way to the Kenyan border. Square lines of fields dotted with shiny tin roofs replace the wild richness of the Serengeti. There live most of the poachers; there the deepest resentment simmers. Our lions wander as far as you can see in any direction, but those that wander up to the north and west may get shot, snared, or poisoned. At times I can convince myself that the Serengeti is a local version of infinity, but from a continental perspective it is all too small.

We don't usually work within twenty miles of the park's western edge, but late last year I brought out a hundred students from the Tanzanian College of Wildlife Management (Mweka) to discover how many lions live throughout the entire Serengeti. Surveying lions over such a vast area is not easy. They are extremely difficult to spot, even when we radio-track them from the air. So we didn't go out to *look* for lions, we went out to listen. I divided the Mweka students into seventeen teams, rented a fleet of vehicles, and sent sixteen of them to every other part of the park. By listening for our study prides around SRI, the seventeenth team provided the baseline, the standard of comparison.

During the day, we all drove systematically around our listening posts and counted prey animals. Silent nights might mean that there are no lions, or that there is no food around. We were lucky. The pattern of rainfall that year had led to a remarkably even distribution of wildebeest and zebra throughout the park. The lions should have been equally vocal everywhere.

For three straight nights we stayed up until midnight, listening for roars. Each small group huddled around its campfire, telling stories of home—Tanzania, Kenya, Zaire, Angola, Zambia, Uganda, Zimbabwe—and trying hard to stay awake. On the second night we were battered by a colossal thunderstorm, and all other sounds were blocked out by the raindrops crashing onto our tents. But in the cold stillness after the storm, invigorated lions all over the Serengeti reestablished contact, telling themselves and us that the phone lines were back in order and that everyone was still there.

We discovered that the Serengeti contains an unspoiled area large enough to hold about three thousand lions. But where lions come up against too many people, the lions always lose. All through the heart of the park, all the way to the eastern boundary, we heard just as many lions roaring each night as around SRI. But along the western agricultural boundary, from the Kenyan border half way down to Seronera, there were no roars at all.

The Mweka students had no trouble discovering why lions were missing from one particular area. A lion had killed two children in a small farming village just a few hundred yards outside the park boundary. The family slept in the same hut as their livestock, and the lion crept in through the front door. The villagers retaliated by setting out poisoned carcasses.

Agriculturists have every reason to feel hostile toward lions and any other animals that come crashing into their lives. Wildlife has no respect for arbitrary park boundaries. Lions eat children, elephants knock down houses, wildebeest trample crops. These people are not going to move away. They have to earn a living somehow; they want to get by as best as they can. Try transplanting this situation to California or Italy, and there is no question how the conflict would be resolved.

But lions do not suffer only at the hands of the local people. Several teams of Mweka students found firsthand evidence of excessive hunting by foreign hunters. A tourist hunter will pay five thousand dollars for a license to shoot a male lion, but in order to qualify for a license he must also pay twenty or thirty thousand dollars for a two- to four-week safari. Hunters bring in more foreign exchange than tourism, and with such large sums of money at stake there is considerable scope for corruption.

The national park is encircled by a ring of game reserves and game-controlled areas where licensed trophy hunting is overseen by a small number of hunting firms. There are strict quotas on the number of animals that can be taken within each concession, but if a rich tourist bags a weedy specimen on his first day out hunting, he will be tempted to find a larger animal later. If he succeeds in improving on his first kill, then the lesser trophy can always be discreetly dumped. The annual quota may be only four male lions, but some firms may shoot as many as fifty.

In most other animal species, hunters can rightly claim that shooting adult males has relatively little impact on population dynamics. After all, younger males are always out there waiting to take over from the fallen breeder. But shooting a male lion is not so simple: kill one resident male, and the surviving members of his coalition may be less able to withstand the challenges of an invading coalition. Once the resident coalition is ousted, their cubs will all be killed. Shoot one of the replacement males, and their cubs may also be lost.

The actual effect of trophy hunting is hard to measure with any accuracy. Lions living in game reserves are very shy, and we can't be sure whether the hunters take mostly resident or nomadic males. If the hunters chiefly shoot nomads, the number of challenges to resident males might actually be reduced, and the fathers might hold on long enough to raise many more offspring.

But we haven't a clue whether they take more residents than nomads, and we can't even trust them to tell us how many animals they shoot in a year, or where they kill them. During our survey, one Mweka team visited a hunters' camp and found the bodies of three lions that had been shot the day before. We had already received reports that this outfit was exceeding its quota by a wide margin; their allotment for the entire year was only five animals. Twenty miles to the south, a second Mweka team found a leopard bait just inside the park boundary. To hunt leopard, you regularly hang chunks of meat in a tall tree. Then, when your client arrives, you take him to the feeding station and wait. It makes a picturesque shot, but there's not much sport involved.

In fact, lion hunting is not much sport either. I've heard of hunters who were so drunk that they could hardly hold their guns. Their guides would drive to within fifteen yards of an animal that would just calmly watch the drunkard take aim. In Texas, you can go to a small island to shoot tame lions that were once house pets. In Germany, "hunters" pay to shoot the thawed carcasses of captive lions that have been medically euthanized and stored in meat lockers.

I may not like trophy hunting, but I have to accept that lions are a renewable resource, a cash crop. And to be fair, not all hunters are mad, bad, and irresponsible. One conscientious group called Tanzanian Game Trackers is run by a group of ardent conservationists. TGT encourages local people to turn in snares and even sets aside some of its hunting areas for photographic safaris. Poachers take far more animals from the Serengeti than tourist hunters. Eliminating poachers would greatly increase the number of animals in the ecosystem, and turning part of a hunting area into a photographic zone has the same effect as increasing the size of the national park.

As a result of our Mweka survey, the Tanzanian Game Department banned lion and leopard hunting in most of the Serengeti reserves for one year, but there is no guarantee that the quotas will be enforced any more rigorously in the future than they have been in the past.

We have turned our backs to the constant menace of the outside world and started to head for home. The turbulent updrafts rising off Nyaraswiga Hill give us one last thrill as we bank slowly around and begin our final descent to the grassy airstrip. The ground slowly comes up to meet us, and the inside of the plane becomes stiflingly hot. Sebastian opens his window, and we taxi back toward the hangar with the sputtering of the engine reverberating inside the plane. The passengers in the backseat are both looking quite relaxed, but Pam is clutching a small white bag.

On the ground, trying not to look relieved, I wonder how Pam and Sarah will cope with the monthly flights for the next three years. Maybe they'll start to crave this sort of visceral adventure. Half the people I know out here are addicted to adrenaline. Take Alan Root, for example, in whose house Pam and Sarah are currently staying. Alan revolutionized wildlife filmmaking with his exhilarating aerial photography and spectacular close-ups of dangerous animals. But Alan has lost a finger to snakebite, been nipped in the butt by a leopard, and nearly lost a leg to a hippo. He came to Catherine's third birthday party with a large bandage around his thigh—he had just been bitten by a gorilla in Zaire. Alan's only regret was that he had missed the shot.

Sebastian invites us to breakfast, and we all gladly accept. Pam lost her first breakfast over the Kibumbus, and I wouldn't let anyone have coffee before takeoff. Before we left home this morning, I warned everyone about the dangers of a full, airborne bladder with one of Alan Root's stories about the first scientific director of SRI. The director was a very proper Brit, a colonial stalwart who always wore starched shorts and white kneesocks and had an extremely precise manner that Alan can mimic very well. He often had to fly on business from the Serengeti in a small two-seater called a Super Cub.

The Super Cub can take off and land almost anywhere, but while it is ideal for floating slowly above the migratory herds with a camera and a counter, it is painfully inadequate for long-distance travel. One day, the director had to fly to Dar es Salaam to meet with several important officials. About midway between Arusha and Dar, his bladder was bursting and he could see nowhere to land. He decided to pee into his shoe and planned to empty the contents out the window of the plane.

Having filled his shoe to the brim, he opened the window, but the wind was much stronger than he had bargained for, and the pee came splashing back inside and drenched him completely. In his shock, he lost his grip on his shoe and dropped it out the window.

He arrived in Dar an hour later. Soaking wet and stinking of urine, he

Wildebeest trekking through Barafu kopjes.

Lion eyes.

(previous page) **Serengeti sunset.**

Serengeti cubs.

Thomson's gazelle crossing the track to Gol kopjes.

A Serengeti male.

An infant chimp.

(facing page) The upper slopes of Gombe.

A young Gombe baboon.

An adult chimpanzee grooms himself.

Infant baboons at play.

A female lion in Ngorongoro Crater.

(following page) **A male chimpanzee looks out over Gombe National Park.**

limped out of the plane on one shod foot and calmly proceeded to introduce himself to the waiting line of dignitaries. "I'm so very pleased to meet you."

FRIDAY, 8 NOVEMBER

The land swells and falls as we drive through open plains that could be in deepest Kansas or Nebraska, except for the distant acacia trees that bob into view as we rise over the crest of each ridge top. It is nearly noon. The compass needle on the dashboard spins around uselessly as we bounce along the open ground.

To our right, the male ego is on full display. A large male bustard struts to and fro on top of a low hill. With tail pressed up flat against his back, wings held down to the ground, head tucked down on his shoulders, and snowy white neck puffed out like a lace ball, he awaits some lucky female.

Over the shoulder of the hill, the Sametu kopjes rise into view. The largest kopje bulges organically out of the ground, pachydermal, pregnant, prodigious, a bastion of permanence in a land of restless change. Atop the boulders rests a fig tree the size of a hot-air balloon. We park beside a shallow streambed that drains a triangular green marsh beside a brackish pond. A small flock of shorebirds paces ceaselessly around the edge of the coffee-colored water.

Shiny black flakes of obsidian and coarse fragments of white quartzite litter the bare ground. These stones do not occur naturally in this part of the park; they were brought here by ancient inhabitants. Dig down around some of these kopjes, and you will find artifacts well over a hundred thousand years old. We're not that far from Olduvai, after all. The Serengeti has been home to humans for millions of years. Australopithecines may have clambered over these very same rocks.

We assemble lunch from the back of the car and scramble up the kopje, clapping our hands and making loud confident noises to shoo away any lions that might be snoozing in the shade of the overhanging tree. At the top, we sit comfortably on square slabs of stone that have spalled off from the underlying rock.

Kopjes were formed deep beneath the ground hundreds of millions of years ago. As the floor of the Rift Valley sank deeper and deeper, the

Serengeti was continuously uplifted. Constant erosion eventually exposed rocks that were much harder than the surrounding formations. Released from all that pressure, they crack like an onion as do the great granite domes of Yosemite.

These stone slabs have served as benches for a very long time. A series of small depressions has been carved in two rows on the surface, like an egg tray. This is a game board, and the game consists of moving pebbles from one pot to the next, trying to capture your opponent's pieces. Played all over Africa, the game is simply called *bao* (board) in more sylvan habitats. Bao boards adorn the top of virtually every Serengeti kopje with a good view, carved out by the Masai some time during the last thousand years.

A few large agama lizards have emerged from the cracks to bask and look for bugs. Rock agamas sometimes act like cattle egrets, hopping on sleeping lions' backs and snagging the idly buzzing flies. The male agama is a garish combination of coral pink and royal blue; the females are dull brown. Their heads bob up and down as they wait for us to attract flies for their supper.

Scattered around our stone benches lie half a dozen piles of dried lion dung—tufts of hair and splinters of bone, the remains of last month's dinner. The soft trunk of the fig tree is etched deeply by a series of narrow grooves running down to the ground. The tree is the Sametu pride's scratching post. We are at the very center of their range, and this is the only large tree within miles. The lions come up here for the shade, the view, and the shelter. The nooks and crannies between boulders are good places to hide tiny cubs.

About three-quarters of a mile to the north, I can just make out an odd object sitting next to a small bush. Borrowing Sarah's binoculars, I can see that it is a female lion looking rather anxiously toward us. We take our time over the rest of lunch, but eat quietly to avoid disturbing her further.

West, past the marsh, there is little to distract the eye. A neat line of kopjes is dwarfed by the platinum blond plains that stretch to the horizon in every direction. No entangling thorn trees, no clutter, no mess. The very simplicity that renders the plains so inhospitable from the air has precisely the opposite effect from the ground. From here, the rolling repetitive landscape provides a deep sense of belonging to a gentle world that recedes away forever, inviting you on and on.

Lunch finished, we drive over to the lion and discover that she is one of our few solitary females, SBG. She was born in the Simba kopjes and still centers her life over there, but she wanders far and wide. I have spent many days with SBG and have probably learned more from her than from any other lion.

Living alone is a tragic fate for a female lion. Solitary females almost never succeed in raising any offspring and often become nervous and furtive. However, they are not aberrant misfits, but perfectly healthy animals that have ended up alone through forces beyond their control.

Young females that are evicted at a male takeover sometimes succeed in establishing a new pride, rearing their own offspring, and settling into a permanent range close to their mothers. However, if there is only one female in the cohort, like SBG, she must make it on her own. Even a female that has been recruited into her mother's pride may end up alone. The Boma mother whose cubs were killed by Smith had once belonged to a pride of four females, but all of her pridemates died within a year of each other.

Once a female becomes solitary, she can only live in a group again if she rears a daughter. Unlike single males, who readily join unrelated companions, solitary females almost never team up with females from other prides. And since they only rarely rear any offspring at all, solitary life is essentially a dead end.

SBG is nearly ten, and she has reared only one cub, a son, past the age of two. He is not with her today, so he has probably died. She will be lucky to live another two or three years and by that time her fertility will begin to diminish. She will have no family to mark her passing.

It should come as no great surprise that solitary females suffer lower reproductive success than females that live in groups. Lions are the only feline species in which the females are so militantly sociable. Male lions only form groups because they have to secure and defend groups of females; male coalitions are largely formed by groups of mothers that have reared their young in a crèche. The fundamental reason why lions live in groups lies with the females.

But what are the precise advantages of female sociality? Why do solitary females fare so poorly?

Until Anne and I studied communal cub rearing, we thought, like everyone else, that lions must surely live in groups to be able to capture large prey. The key to lion sociality rests with the females, the hunters. Cooperative hunting provides such a photogenic notion that it must surely be the answer—the final moments in the life of a zebra collapsing under the sheer weight of all that corporate hunger. The yellow ring of bloodied faces encircling the dismembered carcass. We've all seen it on TV. Why question something that everybody knows to be true?

We certainly *wanted* to believe that everyone was right. We knew very well that a detailed study of hunting behavior would be extremely difficult, and we were never keen to work at night. But strong belief does not advance the frontiers of knowledge, and there was surprisingly little solid evidence that group foraging was indeed advantageous.

In fact, no one had attempted to study group hunting in any detail until George Schaller founded the Serengeti lion project in 1966. George discovered that solo hunters were successful in only one out of every six hunting attempts, whereas pairs were successful about one time out of three. In addition, groups occasionally caught several prey in the same hunt whereas singletons only caught one prey at a time. Curiously, though, George found that all larger groups had about the same hunting success as pairs.

Schaller published his findings in 1972 in an outstanding book entitled *The Serengeti Lion.* In the following years, his data were used by several scientists to estimate the amount of meat acquired by lions foraging in groups of different sizes. Their calculations suggested that pairs of lions gained the most food each day. Although a successful pair would always have to divide its spoils in two, pairs succeeded twice as often as singletons, and sometimes caught multiple prey. However, adding a third member to the group would only add an extra mouth to feed without raising the group's hunting success. The most efficient group size for hunting therefore appeared to be two.

Admittedly, these analyses did not provide overwhelming support for the traditional view, but they at least suggested some role for cooperative hunting. The optimal group size was apparently greater than one, and this is the essential difference between a lion and, say, a tiger. As time went by, though, Anne and I began to question the significance of even this small effect. The vast majority of females live in prides of three or more females, and although pridemates are usually scattered throughout the pride's range, they generally forage in groups much larger than two.

The more we thought about group hunting, the more flaws we found in the whole story. The indirect estimates of feeding efficiency were much too crude. All lions had been assumed to hunt the same number of times each day and to seek prey of the same size. A solitary female may be less likely to succeed when she hunts, but does she hunt just as often as a pair? Surely she would try to compensate for her lower success rate by hunting more often. Large groups may have the same hunting success as a pair, but only a large pride would have the collective strength to pull down a meal as large as a giraffe or Cape buffalo.

We were also nagged by doubts about whether lions actually do hunt cooperatively. Watch a group of hunting lions, and you don't often get the impression of a well-oiled machine. Time and again, one lion will spot the prey, get up, slink off, and position herself in the tall grass. Another lion will then stand up in the open, yawn, stretch, and gape at the now-alerted prey. The prey bounds off to safety, and the lone hunter walks back to her companions as if nothing had happened. No anger, no pep talk: "Now girls, let's remember to work together next time." Just flop and back to sleep.

It seemed to us that the lions were probably grouping together for some reason other than hunting. They all had to eat some time, and if they were together anyway, they might as well hunt together or at least be on hand for the feast should someone else feel moved to exert herself.

We could only resolve the issue by directly measuring feeding rates ourselves. No one had ever made these measurements before, and for a good reason. Lions feed only once every few days, and they usually hunt at night. If the group is large and very hungry, the carcass may disappear within a few minutes. Thus we couldn't just check up on a pride every now and then to see how it was doing. To find out precisely how much meat females acquired when they hunted alone or in a group, we would have to watch them constantly for several days in a row.

Having just given birth to our daughter Catherine, Anne couldn't join me for such long periods of observation, so we recruited Dave Scheel, our first Minnesota graduate student. Dave may have reacted colorfully to the unexpected motions of a light aircraft, but he was steady as a rock when it came to long nights of lion watching.

He and I eventually settled on a standard routine. We would pick a pride with a radio-collared female and watch the group day and night for four straight days just before or after the full moon. We alternated six-hour shifts: I watched each night from sunset until one in the morning, then took over again at sunrise.

Our night-vision goggles turned each moonlit night into an eerie green dreamworld, bright as day but cold, grainy, and indefinite. Lions became rocks became blurs in the grass. Cramped and half-dazed in the darkness of the car, we were pelted by the incessant beep, beep, beep from the transmitter, waiting for a slight change in volume that would tell us when the lion had moved. The blur has a head! She's sitting up now! She might actually do something!

We completed our first few "four-day follows" in the same car, but there was no way to sleep through an active night, no way to get out of the car and stretch. After I acquired a second car for the study, Dave would drive up at the end of my shift, and I would pass him the equipment through the window and bring him up to date. I could then escape to sleep, walk, or even drive back to SRI to visit the family. It felt like being on parole.

I must admit that I still find it difficult to talk about this study. Each night alone in the car consisted largely of just trying to stay awake, trying to stay sane, coping with the endless, mindless soliloquy:

"What time is it? It is 10:08 PM; two minutes later than the last time you looked, three more hours to go.

"What are they doing now? Have they moved?

"Are you kidding?

"Go ahead, sleep your lives away. See if I care.

"It's cold again tonight, shut the window . . . No, better not; someone might roar . . . My foot's going to sleep . . . Stand outside for a minute . . . It's freezing out here.

"Where's the fast-forward button on these night-vision goggles?"

Beep beep beep BEEP.

"Action at last, which way are we going?

"North . . . Plod plod . . . We don't seem to be in any hurry to get anywhere . . . Stay in low gear . . . That's close enough . . . Now stop.

"Is she stalking something?

"Nope, just trekking somewhere.

"Stay back, keep her in view . . . Where are they going? Are they still all together?

"Two, three, four . . . Yeah, everyone's still here . . . This river must be the Ngare Nanyuki . . . I can't cross this . . . What about here? Is this the top of a cliff?

"No, it's okay . . . Go as fast as you can through the mud; slow down and you'll get stuck.

"Now where are they? The signal's loudest off that way.

"There she is.

"Now, where am I?"

At a quarter to one, chilled to the bone, listening to the crackling of the two-way radio, the endless garbled chatter spilling over from a nearby frequency, ascending whistles from a wartime wireless, occasional bursts of Morse code, I finally hear a distorted voice:

"Hello, Craig."

"Hello, Dave. You sound like Donald Duck. Listen, we've come about six kilometers to the north. They've just crossed the Ngare Nanyuki, but I've found an okay crossing. Drive due north about five kilometers, then try tracking us. Flash your lights when you reach the river, and I'll guide you across."

We had to abandon several follows when one of the cars broke down or someone fell in a hole. Dave even came down with malaria one night. But we almost always managed to stay with the lions for the full four days. Success carries its price, however; it took me nearly a week to recover from each follow. Months with two successful follows were especially bad. To this day, I can rarely sleep longer than six hours at a stretch.

But these were the least of our worries. What if we couldn't get enough data to answer our research questions? Was four days long enough to measure each group's performance? Would we find some overall pattern? We had no idea whether we would learn anything at all from our endless hard work.

We completed thirty-six follows in twenty-four months. But by the last half dozen I had to find someone else to go out with Dave. I couldn't stand the tension and boredom any longer. For hours on end there was nothing. The lions slept, slept in the darkness, slept in the heat of the day. The lions slept, and we had to endure the passage of all that time. Month after month, moon after moon. Thirty-six times in all; one hundred and forty-four nights. It has been nearly four years since we finished this study, and I still panic occasionally at the sight of a full moon.

In the end, though, it was worth it, for we did find a pattern after all. Some groups fed consistently better than others, but only in the months when the migratory herds were not around. When the migrants are abundant, the lions can always eat their fill—when a hundred thousand wildebeest wander through your territory, the pickings are easy. But once the herds move off, the lions have to survive on more difficult prey species, and they have to work hard to find enough to eat. These are the times when grouping patterns make a difference to feeding rates, but not in the manner that we had expected.

We discovered two peaks in feeding success during hard times, one for solitary females and the other for groups of five to six females. When prey was scarce, warthogs were the most important prey for all the smaller groups of lions. Pigs weigh only about fifty to eighty pounds, and solitaries can catch them as easily as any pride. During each day of our study, we watched solitaries procure the same total amount of meat as a pair, trio, or quartet—but the solitaries didn't have to share their meals with anyone. Thus, a solitary ate far more than a female in a small group. But groups of five or six females managed to eat as well as solitaries, and they did so by specializing on Cape buffalo.

Buffalo weigh about a thousand pounds, and they are very aggressive, dangerous animals. Buffalo have killed a number of our lions over the past twenty years, and they gore several rangers and children each year. Our own two children are more frightened of buffalo than of lions. Small groups of female lions also seem scared of buffalo, and they won't even try to pull one down. By contrast, very large prides seek them out, but they attack with great care, encircling the beast, jumping on its back, and steering clear of its thrashing horns. A successful buffalo hunt requires coordinated cooperation and a division of labor.

So if you are a female lion, the migratory herds are out of reach, and you are hungry, you have two options: split off from the rest of your family and go catch a pig, or, if your family is large enough, stay together and brave a buffalo. The two options will provide you with about the same amount of meat each day. Foraging alone is no hardship, but foraging together is unrewarding unless you have enough firepower to catch a dangerous dinner.

Our findings enabled us to rule out cooperative hunting as the explanation for female sociality. Even for females in large prides, catching a buffalo only allowed them to feed as well as a solitary, but no better. Females in groups of intermediate size fed so poorly compared to solitaries that you might expect females in small prides to spend long periods alone. However, except when they are tending newborn cubs, females hate to be alone for very long. Females in small prides may forage by themselves for brief periods each week, but they spend the rest of their time in each other's company, feeling hungry.

Before our study, the typical view of lions was that of a solitary species that came together to hunt. Although pridemates may sometimes assemble for a buffalo hunt, lions should be viewed the other way around: a social species that occasionally has to split up to hunt.

But if cooperative hunting is not the answer to female sociality, what is? Crèches certainly function to protect cubs from infanticidal males, and mothers are the most gregarious females in the pride, but that's not the whole story. Childless females can also be quite sociable, especially in small prides where group hunting is particularly disadvantageous. If infanticide were the only force driving the evolution of lion sociality, we might also expect to find a number of other sociable cats: male cougars, leopards, and tigers are also infanticidal, yet the females of these species are all militantly solitary.

Although our four-day follows were originally designed to test the cooperative hunting hypothesis, our time spent with solitaries like SBG gave us important insights into the overwhelming value of group living. Solitary females may get plenty to eat, but their world is filled with terror.

We first followed SBG during the height of the dry season. The plains were virtually empty, and she moved nearly thirty miles in four days, passing through the ranges of several other prides. Just before midnight on the second night, she came upon a lone zebra, a zebra that didn't seem to see well even in the bright moonlight. She stalked it for only a few minutes, creeping forward through the short grass whenever the zebra grazed, freezing when it looked up.

The zebra finally sensed something was coming and started to stumble off, but SBG covered the final few yards in only a few bounds, leapt up, clamped onto its throat, and pulled it down. The zebra died in silence. SBG had captured her photogenic prey without any help from anyone.

After resting briefly, she opened its abdomen and fed, ripping out entrails and wolfing them down. Then she froze completely. I stuck my head out the car window, but I couldn't hear anything. SBG obviously heard something, though. She started to feed even faster, on liver, spleen, all the soft parts.

She suddenly froze again, staring into the night, and two hyenas wandered up from the darkness. SBG stood over the carcass and frantically tore at the zebra's guts. The hyenas kept a nervous distance, but then one started to whoop, a haunting call as characteristic of the African night as the lion's roar.

SBG really started to panic, not in fear of the hyenas, but of whatever the hyena's calls might attract. SBG had strayed well into the heart of another pride's range, and she did not want to be surprised by a resident pride claiming its territorial rights. Both hyenas called in unison, and SBG stared off in the same direction one last time before fleeing into a nearby kopje. Just moments later, a large male lion charged up and scattered the hyenas from the eviscerated zebra. SBG must have heard him coming from nearly a mile away. She stayed hidden in the kopje for the next six hours, then left cautiously just before dawn. She had slipped in and out of the strangers' territory, quiet as a mouse.

The following year we watched SBG at the height of the wet season. Thousands of wildebeest and zebra darkened the plains around her home at the Simba kopjes. At night the wildebeest clumped together and drifted around like huge black rafts.

SBG never bothered to hunt. She just listened for the grotesque cackles and giggles of hyenas squabbling at a kill, then charged in, chased them away, and fed to her heart's content. With so much food around, she hardly needed to move an inch, but she roamed around her territory each night, roaring: uuuunnnnhhh uuuunnnnhhh uuuunnnnhhh uuuunnnnhhh unh unh unh unh unh. This place is mine. Mine. Mine. Mine. Mine.

On the third night, SBG walked toward the three scattered kopjes that marked the western edge of her territory. She suddenly stopped, crouched, then turned tail and crept off, waddling low to the ground and moving fast. A minute later three female lions walked right past the car. The invaders sat down, and four more females arrived—the seven females of the Plains pride, patrolling the southeastern limit of their territory.

Hopelessly outnumbered, SBG fled to a ridge top about a mile and a half away. She lay quietly by a shallow waterhole for nearly an hour, then started to roar and roar and roar. She must have roared twenty times that night. But it was an exercise in utter futility. She could only hope to defend her territory against other solitaries. Any group of lions could chase her away.

A solitary female has to survive in a world bristling with land-hungry prides. SBG has only been able to carve out a marginal home at the edge of the world. Her territory contains a half dozen barren kopjes, a few resident prey, and no permanent water supply. It is territory that no pride would want. Large prides control much richer property—stretches of river in the woodlands, or large clusters of kopjes near a spring or pond, places where females can keep their cubs hidden in close proximity to a regular food supply. SBG could never have fed small cubs from her long-distance zebra kill—she had caught it nearly twenty miles from home.

A good territory is essential to successful reproduction, but good land is the object of everyone's desire. It must be defended forever. When George Schaller started the lion study in 1966, he focused on three prides along the Seronera River: the Masai, Seronera, and Loliondo prides. In the beginning, all three prides contained large numbers of females, and they all held rich territories, but their subsequent histories diverged as dramatically as those of any three countries in the Middle East.

The Masai pride still lives along the Seronera River in the northern part of their original range. Over the years, two large cohorts of their daughters established new prides, the Masai Kopje and Sametu prides. These three prides subdivided the original range, together occupying no more or less than had their great-great-grandmothers. The big old house has been broken up to shelter each branch of the family.

In contrast, the Seronera pride showed a constant pattern of division and fragmentation. By the time Anne and I arrived in 1978, the descendants of the original pride had split into three small prides and two solitaries. Once the size of each group fell below three females, the surviving pair became the target of frequent incursions by their neighbors. Inevitably, each small fragment dwindled to a lone female who quickly lost her share of the original territory. Today there are no female descendants of the Seronera pride. Divided we fall.

Meanwhile, the Loliondo pride built an empire. Year after year they produced large cohorts of daughters. Some remained with their mothers, and the rest established new prides nearby. Descendants of the Loliondoes can now be found in five prides. One of these, the Campsite pride, annihilated the last fragments of the Seronera pride, and handed on the remnants of the Seroneras' old range to their dispersing daughters. The Loliondo nation has retained all of its original territory and acquired most of the Seronera pride range as well. To the victors the spoils.

The traditional idea that these great beasts aspired collectively against the elements was somehow ennobling. The cooperative hunters were grand and generous, free from the flaws that mark our own species. However, the real answer shows that they, too, have a darker side. Group territoriality is the preeminent cause of sociality in lions. If a female loses her own territory, she loses all hope of raising cubs. If she is discovered in another pride's territory, she may well be killed. Her chances of maintaining her territory depend on the size of her pride. Once a pride contains fewer than three females, it will almost certainly become extinct, exterminated by its neighbors.

If you are a female lion, your whole life depends on the number of companions you can muster for the next showdown. It is high noon, the train is coming. The gang of four is on its way. How many partners do you have? If you are in a small pride, you need to stay together as much as you can, ever ready for the next battle. If you are in a large pride, you can afford to spread out a bit more. Part of your family will be together at the crèche or preparing to hunt a buffalo. A chorus of roars each night will inspire terror in smaller neighboring prides.

But hold on—lions can't count, can they?

They most certainly can count, and they count roars, too. Karen McComb studied female territoriality by using playback experiments in the same way as Jon Grinnell. When Karen played female roars to groups of different sizes, the speed with which the females arrived at the speaker depended on the odds. Play the roar of a lone female to a group of females, and a trio approaches the speaker much more quickly than a pair. Play the roars of three females to a trio, and they take as long to arrive at the speaker as a single female hearing another solitary. If the females are outnumbered, they don't approach the speaker at all.

Karen also discovered that females that had been temporarily separated from the rest of their pride would often roar in response to the recording, presumably to enlist help in evicting the recorded intruder. Big numbers are advantageous only if everyone is together. And in fact, pridemates often showed up within minutes of the female's roar—comrades in arms, side by side, ready for the next sign of trouble.

But what makes lions so special? Why should lions defend territories in groups when members of every other cat species defend their territories solo?

Start with the lions' diet. Wherever they live, lions specialize on the species of prey that are most abundant. In the Serengeti they prefer wildebeest and zebra, species that form the largest herds on earth. When the wildebeest and zebra are not around, the lions survive on prey that range in size from warthog to buffalo. In some years—the vintage years—favorable rainfall patterns keep the migratory herds in their range for most of the year, and the lions rear many cubs.

Although lions cannot always breed successfully during harsh years, the adults can easily survive. Lions are large animals, and their size enables them to fast for weeks—large species can survive longer on a given percentage of body fat because of their lower metabolic rate. And if life really gets tough, the Serengeti lions can always trek for a day or two to the migratory herds.

Consequently, lions live in high-density populations and are the most abundant of all the big cats. Throughout their range, lions are typically five to ten times more common than leopards, cheetahs, or tigers. We know from studies of other animal species that individuals living at very high densities will more often defend their territories jointly. The more densely populated an area, the more difficult it is to keep out intruders by yourself; groups can patrol a territory more effectively. Female lions are surrounded by competitors, each one seeking the best place to rear her cubs, the best place to find a regular meal. The enemy—other lions, other families—is everywhere.

In a group, a family of lions can cordon off an area large enough to contain adequate food all year round. When possible, a family annexes adjacent land to help its daughters start a new home next door. Females live in groups in order to gain an advantage over solitaries. But once one family starts to live together, so must every other lion. If you live in a group, I have to live in a group, too. If your group gets larger, then so must mine.

Only a few other cat species ever attain the same densities as lions. Bobcats and servals, for example, are quite common in some areas. But these species eat mice and lizards. If they foraged together they would have a very difficult time getting enough to eat.

I've spent much of the past ten years trying to puzzle this out. I can outline my grand theory of lion sociality, but I know it is incomplete; there are still a number of holes in the story, a number of inconsistencies. But this is science. You start out knowing nothing, and you acquire knowledge only slowly. Progress seems rapid at first, but then you realize just how much you don't know. New questions arise and old answers start to fade.

I do think we've made a reasonably good start to the story; territoriality does seem to be the key. Groups of females will kill strangers in defense of their territories. Solitaries either have lousy territories, like SBG, or lose them altogether when their pride shrinks to just one or two females, like the Seroneras. I'm also reasonably confident about the hunting data. All group sizes must surely feed equally well when the migratory herds are around; there's always some sick or lame animal nearby, waiting to be eaten. The dry season data make sense, too. We had already suspected that females would sneak off alone to fatten up during hard times. Female trios really do refuse to tackle buffalo, and suffer from a low food intake. However, trios somehow manage to raise as many surviving offspring as prides that can catch buffalo. Either food intake has very little effect on cub survival, one of our measures is wrong, or something else is going on.

We do know that some other effect of grouping must be penalizing larger prides. Per capita, females in large prides reproduce as poorly as solitaries, and large prides frequently split in two. Perhaps these females are hampered by their prior success, rearing only a few cubs in one year because they had raised so many daughters a few years earlier. Their older daughters may have established a new pride next door and started squeezing the mothers out, competing for the same land, the same pool of water. For lions, the conflict never ends: the configuration of territories that we see today reflects an endless past fractioning of clans, the balkanization of the Serengeti.

So this is my dilemma; why I didn't want to leave home this year, yet why I had to come. Sitting in my computer are the data I need to sort out the next part of the story, the information to measure the costs of factional warfare, to reconcile the disparate strands between measurements of success.

We are bound to uncover more and more questions, and someone has to keep constant track of these lions if we are ever going to answer them all. We can't just drop everything, stare into our computer screens for five years, and then come back refreshed and ready to stay up all night for a few more months.

If I had arrived for the first time today, I would have come out here and found a single female lion sitting under a bush. Pretty boring. But instead, I saw SBG and all her tragic history. These animals are interesting to us precisely because we know so much about their background.

We will always need to know who is who and where they come from, what happens to each lion, and why. Someone has to record all those tiny whisker spots for each cub, measure the size of each pride's territory, follow the fate of each frightened subadult separated from its mother for the first time.

The Loliondoes are the current champions of the Serengeti, but history never ends. No, it never ends. And someone has to be here to write it all down.

Leaving SBG behind, we head to the Sametu marsh. Today our quest is for water, not lions. I need to show Pam and Sarah the waterholes on the plains, and Christine wants to check them for parasites.

During the dry season, the ground dries up. The streams run dry, and the ponds recede, leaving only one small waterhole in every hundred square miles of dry plain. Some dry-country animals can survive without water, but most species have to drink every few days—all from the same small source. Everyone slobbering in it, standing in it. Some even wallow in it, urinate and defecate in it. Occasionally someone dies in it. All the pestilence-stricken multitudes washing and drinking from the same cup.

The water in the Sametu pond is dark and opaque. Aquatic insects dart up out of the murk, then dive down again. Hoofprints line the shore. White crystals of alkali encrust the cracked, dried mud around the edge. Christine sweeps her net through the water and pulls out a few small, wriggling invertebrates: copepods, fairy shrimp, and mosquito larvae. She transfers her collection to a small jar and fills another with water. The copepods may carry a tapeworm that infects the lions. The water is swarming with unseen horrors.

We trudge through the marsh across a spongy mat of dark green reeds. During the dry season, mothers of the Sametu pride wander for miles over the empty plains, searching for food, and leave their cubs hidden here for days at a time. The reeds provide a safe haven for the defenseless cubs, a labyrinth of narrow passageways beneath an impenetrable roof of thick stems. Small cubs can dive to safety in the tangled mass, out of the reach of hostile adult lions or hungry hyenas.

On the opposite side of the marsh, half a dozen wells have been carved in a ledge of bare rock, bright green circles surrounded by snow-white chalk. Moss floats on the surface, but the water beneath is fresh and clear. Most mammals prefer well water to the bitter pond. But the wells lie within a few feet of the reeds, where the lions might be hiding, so most herbivores drink in the open at the comparative safety of the pond.

Sampling completed, we hop in the car and head farther southeast. We continue beside the twin ruts of the Sametu track, but the track peters out as we approach the next ridge. One rut becomes an animal trail, carries on in the same direction, then disappears entirely.

Pam and Sarah chart our course on a large map while I point out the few landmarks on the plains. The hills of Seronera are twenty miles behind us. Fifty miles in front are the mountains of the Crater highlands. The Gol kopjes are about ten miles to the south.

The grass becomes shorter and shorter until we are finally driving on a well-groomed lawn. Left in peace, the grass here could grow several feet high. But thousands of square miles are grazed to a stubble each year by millions of cordless razors. The grass is enriched with calcium and phosphorous by the fresh volcanic soil, and it is eaten with enthusiasm. I wonder how it tastes to the wildebeest. What would make them travel so far, make them eat every last blade? Is it like French food after a year of porridge? Or like a trip to the vitamin cabinet? Is it sweet, or is it sharp and savory?

The ridge top is rough, dotted with pig holes and rabbit burrows, but at least we can see where we are going. Up ahead, dozens of zebra and topi are gathered around a limestone pan, but they scatter in all directions as we drive up. A small freshwater pond, about thirty feet across, has been churned chalky brown by hundreds of hooves. Thousands of tiny organisms dance in the muck. Scattered rainfall hit the bull's-eye here, enough to turn the land green for a few miles in each direction, but only for another few days. The wind is blowing hard across the shallow water; the waves lap against the pitted shore.

During the wet season, dozens of these ponds dot the plains, each one complete with its own waterborne community. Terrapins are the largest residents, and they patrol their tiny oceans like submarines, searching for thirsty birds. They can submerge themselves completely in water only a few inches deep, then up periscope: an eye spots a sand grouse drinking on the far bank. Down periscope without a ripple. The grouse cautiously moves a bit further into the water, eager to bathe. It starts to squat down, and then panic. Wings flap madly, and the bird is airborne. The wingbeats part the water to reveal the terrapin's gaping mouth, its dark green shell.

Late one evening, I watched a pair of Egyptian geese swim over to a small flat stone jutting out of the water. The sun had just set, and the geese were ready for bed; the Sametu pride was still fast asleep at the water's edge. The geese were faced with an interesting dilemma: the stone was only large enough for one of them, and its mate had to stand in the water.

Once settled, husband and wife tucked their beaks under their coiled necks and went to sleep. Suddenly, the goose in the water started to honk loudly and flapped its wings desperately. The lions sat up and watched as the goose flew up a few inches with a large terrapin clamped firmly to its foot. The bitten goose finally pulled itself free and flew safely to shore. It complained loudly to its mate, but after calming down they both swam back to the stone.

Every few hours, I heard the same honking and flapping as the drama repeated itself in the darkness. I still wonder whether they took turns standing on the rock, or whether one insistent spouse kept flying back to the stone, victimizing its waterlogged mate. Grounds for instant divorce.

Toward the eastern plains, we weave through a series of erosion terraces, shallow steps of bare soil every hundred yards or so that form a random stairway up the next ridge. At the top, we suddenly find ourselves free from all bumps, ruts, and holes. Now the driving is like flying—we can go in any direction at any speed. There is no sense of being stuck to the ground. The plains rise and fall almost imperceptibly. The air is so clear that we can see the clouds beyond the horizon, we can see the curve of the earth. Everything sweeps away from you out here, there is an extraordinary sense of freedom.

We roll through the Barafu kopjes, each one different from any we have seen before. The granite is pink with a smooth, rounded, and unbroken surface, like the skin of a subterranean creature whose sinuous body has just thrust through the surface of the earth—a small reminder of a gargantuan struggle beneath us that reduces our own lifetimes to a mere flash of light in a drama of literally earth-shattering significance.

Hundreds of game trails snake out of a flat-bottomed gorge that bisects the constellation of kopjes. The ground is suddenly a washboard, and we have to drive along slowly. Down there somewhere is a stand of fever trees—the yellow bark will tell us where to find open water.

At the spring we find a small earthen dam connected to a cattle trough freshly dug in the ground. Masai come into the Barafu gorge each dry season, though they are supposed to remain outside the park. The eastern park boundary is only two or three miles away, and the park rangers rarely patrol this area. The Masai only come here to water their stock, and their worst offense is to chop thorny branches off the acacia trees to build temporary corrals for their cattle.

More specimen jars filled, we drive up from the gorge and roar along to the south at sixty miles an hour. The compass still spins about, and only memory provides a sense of location. Finally a view, and I'm about two miles too far east, but we've almost reached the Gol kopjes. Here comes Dolphin kopje, the single crescent-shaped dome a beached cetacean on the plains; a fig tree shades its head.

We park in the shade and scramble up the smooth sides of the dolphin's body. Near the top, the infolded rock forms a pouch four or five feet deep that fills whenever it rains. The pool never dries, it's still full today. From the top we can see four scattered kopjes a mile or so apart, each with its own distinctive character. To the south, the yellow plains recede down, down toward the Crater highlands; the trees of Olduvai Gorge draw a penciled moustache on the distant plains. To the east, through a pass in the Gol mountains, stands Oldonyo Lengai, a virtual infant in the ancient landscape.

As we roll over the ridge top to the west, the remaining Gol kopjes come into view. The granite is grey again over here, and we are in the world's largest Japanese garden. Everything has an austere permanence, an elegant simplicity. Every rock might have been placed just so by a gardener with impeccable taste.

The Gol kopjes make an unforgettable impression, and they have obviously implanted themselves in the human mind for unimaginable periods of time. Many of the kopjes are surrounded by large stone circles, man-made, a hundred thousand years old. Others possess a delicate pinnacle or spire that could easily have been toppled by a minor act of vandalism. Over there is a row of monoliths standing in an arc, hooded monks contemplating the cosmos. Most kopjes are shaded by one or two full-sized acacia trees pruned to ideal proportions by the Masai. Easter Island meets Stonehenge in a Shinto shrine, but on a panoramic scale.

At the southern end of the Gols, a local rainstorm has invigorated a large kopje with a display of brightly colored flowers. Squat, red fireball lilies throb violently in a setting of severe grey stone, delicate red and yellow Gloriosa lilies dangle from the branches of a wispy green thorn tree. Orange candelabras sprouting from the aloes attract metallic-green sunbirds. At the top of the dome, weird succulents and sturdy white vines loop down into a circular pool of water, an unexpected scene straight from the rainforests of Indonesia or Sri Lanka. Scattered everywhere are dried lion and hyena turds. Hyenas eat bones, and their turds are almost pure calcium—they look just like golf balls.

Composed of four closely spaced domes, the kopje looks like the print of a lion's paw when viewed from the air. Looking down from the top, we can see that the surrounding grassland is as smooth as a billiard table, but wait—it's not unblemished. Coming directly over from the park gate at Naabi Hill are a series of rutted car tracks. They weren't here a year or so ago. Newly gouged tracks circle the Lion's Paw, leading to each of the other kopjes like spaghetti junction.

For some reason, tourist buses follow each other's tracks, eat each other's dust. Drive over this short grass once and it springs back with the rains. Drive over and over the same spot and the grass is destroyed—you've created two parallel white lines in the dust that may not grow back for years. This simple landscape is not particularly fragile, but mindless tourism can destroy the very essence of the experience.

Blazing our own trail, we head up to the central cluster of the Gols, the high point on the plains, the great divide between the western drainage system that ultimately flows into Lake Victoria and the Nile, and the dead-end system that empties into Olduvai Gorge. Each kopje up here has a shape of its own. One is a simple pyramid of watermelon-sized stones, another is just two massive grey spheres separated by a perfect acacia tree. A third forms a stark granite wall, lined on either side by half a dozen well-spaced trees. At the very top of the ridge stands a balanced rock about eight feet high on a flat anvil of stone, right next to a cubic boulder with rounded corners, shaded on one side by a single acacia. Beyond to the southeast, picture-perfect in the warm light of late afternoon, Ngorongoro and its attendant volcanic peaks rise up to the coloring sky.

On those dreary days in America when I wish with an intense pang that I was in Africa, this is the spot I long for. This is the spot where we always spent Christmas when it was endlessly green and a million wildebeest grazed for miles in every direction. The countless black profiles, their wispy brown beards back-lit in the golden sunlight, grazing on the green-felt landscape, grunting contentedly; the air permeated by a barnyard smell of freshly digested grass. If I want a moment of mystical exaltation, this is the spot that sends me six inches off the ground. This is my place of worship.

Solidly on the ground once more and heading west, we approach the largest kopjes at Gol, large-scale monuments overlooking the long plain down to the main road and beyond to Seronera. Tombs fit for a pharaoh, built by no one. Testimonies to the power of blind, random nature, right next to the spot where a female from the Gol (GO) pride abandoned her ailing cub to the hyenas, the spot where the week-old GOD met his untimely end.

At the north end of the line we find the official track and drive beside it down a succession of erosion terraces, directly toward the sun as it sinks behind one last bank of low clouds. The wind follows us, blowing dust inside the car whenever we slow down for a bump, ease down over a terrace, or bounce through a pig hole.

By the time we rejoin the main road, the car is overheating from the trailing wind and the grass is knee-high. As if we needed any further reminder of the futility of human endeavor, Christine looks in the back of the car to see that her box of two dozen water samples has turned on its side. None of the jars were watertight. The tainted waters from so many sacred sources lie blended in a puddle on the floor.

SATURDAY, 9 NOVEMBER

Pam is at the lunch table, writing letters and singing along with *Carmina Burana:* "Oh, oh, oh. Totus floreo." Sarah studies a local map and enters today's lion sightings in the data file. Christine has gone next door to deal with this morning's fecal samples. Erin has taken Rob to the Kenyan border to catch his flight to Australia.

Wearing a pair of rubber gloves, I reach into a green plastic bucket and pull out a mosquito net that has been soaking in insecticide. We haven't seen any mosquitoes around here yet, but we must prepare for the trip to Gombe. I have to start thinking about baboons and chimpanzees again, as well as the change in climate. It is hard to imagine the damp, musty fug of Gombe while sitting on this verandah, surrounded by the lofty, arid hygiene of the Serengeti, but we must expect the worst. Gombe is one of the unhealthiest places I know.

Nets spread out to dry, I sit down to make a list of equipment for Pam and Sarah to assemble during our absence. After Gombe, we will reunite at Ngorongoro Crater, where we study several other lion prides. We have no housing in the Crater, so Pam and Sarah will need to bring tents, cooking gear, and bedding for a weeklong camping trip.

Christine estimates how many vials she'll need to sample the Gombe baboons and I start to calibrate the GPS equipment for surveying a new Gombe map. Then from somewhere off to the left of the house, we hear a familiar voice:

"*Hodi* (May I come in)?"

To which everyone answers: "*Karibu* (Welcome)!"

It is the old man Kisiri, who has worked at SRI for many years. He is garrulous, colorful, and industrious. Some people find him intensely irritating, but SRI would only be an empty collection of houses without him.

"*Jambo, Bwana Paka! Habari za siku nyingi?* (Hello, Bwana Packer. How have you been for so long?)"

"*Jambo, Bwana Kisiri! Habari yako, mzee?* (Hello, Kisiri! How are you, *mzee?*)"

"*Nzuri sana, ahsante. Habari za mama, na watoto?* (Very well, thank you. How is your family?)"

"*Nzuri sana, ahsante, mzee. Na zako?* (Very well, thank you, *mzee.* And yours?)"

"*Nzuri, ahsante. Habari za Marekani? Engurandi?* (Fine, thank you. How is America? England?)"

"*Safi. Habari za hapa; habari za afya?* (Fine. How are things here? How is your health?)"

"*Nzuri. Na baba? Na mama?* (Good. And your parents?)"

"*Ahsante, mzee. Habari za kazi?* (Thank you, *mzee.* How is your work?)"

"*Wapi? Kazi gani, bwana?* (Where? What work?)"

Greetings finally over, Kisiri reveals the real purpose of his visit. He has fallen on hard times. He is faced with a possible prison sentence.

Kisiri was the headman at the SRI staff village, until recently. As the village elder and a skilled handyman, his duties at the institute consisted largely of repairing the deteriorating buildings and organizing manual laborers to cut the grass and to fill in potholes, keeping up appearances to remind everyone of SRI's former glory.

SRI was founded during the colonial era by John Owens, the last British director of Tanzania National Parks. The institute was developed in the mid-'60s with large grants from several major foundations in Europe and North America. Office buildings, laboratories, and a library were built in the middle of a large kopje, and over a dozen houses were constructed nearby. Scientists came to SRI from all over the world, and a staff of skilled Africans worked as technicians, field assistants, and laborers. Unfortunately, this scheme did not adequately address the aspirations of the Tanzanians, and SRI often gave the appearance of a colonial country club. There was a tennis court and a gliding club, but no Tanzanian scientists.

When Tanzania embarked on full-scale socialism in the mid-'70s, SRI was precipitously nationalized. The British research director (he of the shoe and Super Cub) was fired and hastily replaced by a Tanzanian microbiologist with little interest in wildlife. He was also ill-prepared to sustain a research facility that had been funded by foreign grants. Obtaining support for such a large operation is a full-time job that requires extensive international contacts. Foreign financial support for SRI virtually evaporated overnight.

The Tanzanian director tried to vanish one night, too, having embezzled a large amount of money. He attempted to leave SRI in a light aircraft, but there were no lights on the darkened airstrip and he came to a sudden halt in a tall thorn tree. The plane was ruined, but he escaped unharmed and fled overseas.

SRI was subsequently reorganized, headquarters was moved to Arusha, and the Serengeti staff became increasingly demoralized. Supplies were unobtainable, vehicles were seldom maintained, and snares appeared in the riverbeds. Faced with this sharp decline, the technicians left for well-paid jobs in Arusha, Nairobi, or Dar. Nowadays the Tanzanian staff at SRI consists of an administrator, a secretary, an accountant, a driver or two, several night watchmen, a crew of manual laborers, and a small village overflowing with their extended families. Only a few of the scientists are Tanzanian; most researchers still come from abroad.

Kisiri has worked at SRI for most of the past twenty years, but when life became bleak in the '80s, month after month passed with no real work and no paycheck. Kisiri started asking for odd jobs at about the same time that the water pipeline gave out and a crime wave hit SRI. We hired him to paint our roof, build foundations for our rainbarrels, and reinforce our storage vault. He resurfaced our driveway with gravel and built the lean-to garage, cut the grass, and helped collect firewood.

His arrival every afternoon became a high point in our routine. Proud of his skills, he worked with great competence and good humor. He entertained our children, Catherine and Jonathan, by raising his voice to a falsetto and teaching them to speak in Swahili. They were fascinated by his long, pendulous earlobes that once held tribal ornaments the size of a film cannister.

Kisiri worked like a man possessed. He was a man with a dream, and that dream was to own a bicycle. With a bicycle he could get around his home village in style. With a bicycle he could do business in town. We brought him clothes and watches from America, and occasionally loaned him money. However, we were running out of things for him to do, and he already owed us about a year's wages.

His wife was too *kali* (fierce) and couldn't have any more children. He was approaching sixty—who would look after him in his old age? He must have a new wife. He gave up on the bicycle and married a woman in her early twenties. Her father demanded a bride-price of fifteen cows.

A short time later, a new outbreak of crime hit SRI. Someone had stolen a solar panel from one house; a large bundle of car parts was taken from the garage at the institute. I decided to hire Kisiri as an *askari* (watchman). He could live in the small servants' quarters behind the Lion House and watch over the place when no one was around.

Alas, during his tenure as Lion House watchman, the thefts continued, and equipment was repeatedly stolen from every house at SRI except ours. Kisiri was inevitably accused of being the mastermind, and he was taken off to jail, although the evidence was purely circumstantial. After being beaten by the police, he was eventually taken to court, but his trial was postponed three times. In the meantime, he has been unable to work at SRI and is only allowed to visit for a day or so at a time.

I find it hard to believe that Kisiri would mastermind anything illegal, though he might well have turned a blind eye to burglary. It seems more likely that he was framed; he must have been widely resented by the rest of the staff for his special status in our neighborhood. However, there is no way of teasing the truth from the stories I've heard since I've been back, and, as much as I love the old man, I can't take a public position until after the judge reaches a verdict.

Kisiri is one of the most consistently cheerful people I have ever known, and also one of the hardest workers. He is here today to ask for help, and I want to do what I can. He needs money. He has had no salary for nearly six months, and he has only been able to buy his father-in-law five cows. His new wife wants to leave him, but he doesn't want to lose her. He is terrified of old age.

Our conversation is in Swahili, but I frequently have to stop and look up legal terms in a Swahili-English dictionary.

"Will you be able to come back to SRI if you are acquitted?" I ask.

He looks me straight in the eye. "Yes."

"Can you have the same job back again?"

"Yes."

"When do you go back to court?"

"The first week of December."

"Okay. If you are acquitted, I will talk to my students. If they want to hire you as *askari* again, you can move back in. If you are innocent, I will give you enough money to buy the other ten cows. Come back after the verdict."

Kisiri seems reasonably satisfied under the circumstances, though his brave facade is starting to crack. He is embarrassed and confused by his predicament; he is haggard and for the first time I notice his age. I can also see where his new wife nearly pulled off one of his dangling earlobes; it is wrapped pathetically over the top of his ear. We say farewell, and he stumbles off in his best clothes: blue jeans and a long-sleeved shirt Anne and I brought from Minnesota five years ago.

I can't possibly do more for him now. Many of the local Tanzanians live in such desperation that virtually anyone may have to turn to crime at some point just to eat. After all, Kisiri was my insurance policy against theft, and it paid off for me! Everyone else at SRI should have taken similar precautions. I want to believe in Kisiri's innocence, but in the end it is difficult to trust anyone, a lesson we have had to learn the hard way.

During one of our trips back to the States, we left our Land Rover with another pair of researchers. They lived in an old wooden house about a mile from the Lion House and kept our vehicle parked in the shade of a giant umbrella acacia. One night, after they had gone out, their gas refrigerator sprang a leak. Gas filled their kitchen until the pilot light ignited an almighty explosion. The house burned furiously; gas cylinders and petrol cans went off like rockets.

The conflagration was only a few hundred yards from the SRI staff quarters, and three of the staff saw that the tree above our car had caught fire and would soon collapse on its roof. They tried to push the car out from under the tree, but the engine was in gear and the windows were locked. They broke the rear window, and one man, Barnabas, climbed inside. He took it out of gear, but he couldn't turn the wheel because of the steering-wheel lock. He found a hammer in the tool box and banged at the lock as the flames flew higher and higher. The intense heat blistered and peeled the paint on the side of the car by the fuel tank. Barnabas finally broke the lock and turned the wheel, and the other two men pushed the car to safety.

It was a spontaneous act of courage that saved the lion project: our research support was at a low ebb, and we might not have been able to replace that car.

By the time we returned to the Serengeti, one of the three men had been transferred to Arusha, and a second had resigned. Barnabas had become the chief mechanic at SRI; we treated him as a hero and gave him our absolute trust. But during our next trip to America, someone broke into the Lion House. As best we can tell, the SRI storekeeper found a key that fit our back door. He and another man went to the storage closet in the hallway, broke down the door, and stole two metal trunks. The two men took the trunks outside, then smashed a window to make it look as if they had broken in.

The second thief was rumored to be Barnabas. He was fired shortly thereafter, and we built our vault. Welcome to the fortress.

This is just the sort of story you hear about Africans all the time. It suits the prejudices of the crusty old farts in Surrey or Nairobi who sit around and tell these tales as they sip their whisky and long for the good old colonial days.

But now consider Ephata. Ephata became the SRI accountant shortly after our burglary. I didn't take to him at first. He seemed a bit too smooth, a bit too eager to please. However, late one moonlit night he was summoned to the front door of his house by an urgent knocking. He stuck his head out the door and saw a dark figure holding a *panga* high above his head. Ephata ducked back inside as the three-foot knife sliced down through the air.

Ephata fled to his bedroom and barricaded himself inside. Three men tried to force their way in, demanding the keys to the SRI safe. They wanted to steal the month's wages. Ephata refused, and they repeatedly threatened to kill him, hacking at the bedroom door with their *pangas*. The door was just about to give way when the invaders were startled by an approaching car. They ran off into the darkness, and Ephata spent the rest of the night on the floor under his bed.

The next day Ephata reported the break-in to the police but said that he couldn't recognize any of the three assailants. He didn't even want to talk about it. We were all amazed at his bravery and honesty but surprised to learn that he could not or would not identify the culprits.

That afternoon, a policeman from Seronera came to Ephata's house. He looked around outside and saw the sandal prints of one of the robbers who had tried to decapitate Ephata. Looking closely at the prints, the policeman saw that the sandals were made from tires with an unusual tread. He summoned the SRI staff to line up in front of the house. Once the night watchman realized what was happening, he started trembling. His sandals matched the prints.

The watchman was arrested and taken to the police station next to park headquarters. His brother was a ranger, and the watchman was allowed to stay in his brother's room until he could be taken to a jail outside the park. He was utterly disgraced; he became dejected and refused to talk to anyone.

One afternoon, the brother went out and left the watchman alone with his antipoaching rifle. The watchman picked up the gun, stuck the barrel in his mouth, and blew his brains out.

Ephata is from Arusha and the watchman was from Ikoma, to the west of the park. The two men were from different tribes. Ephata came to me in terror. "Most of the people here are Wa-ikoma. They blame me for the watchman's death. They will kill me."

"But you refused to identify the men who attacked you. The police found him on their own."

"That doesn't matter. I reported the attack to the police. That is why they arrested him. That is why he died. I must leave this place immediately. They will kill me. You know some good people. Help me find another job somewhere else."

Ephata had managed to maintain his composure after the attempted robbery, but now he was quivering and grey. I wrote a hasty note to the owner of a small lodge near Ngorongoro Crater, explaining what had happened and suggesting that if she ever needed a reliable accountant, Ephata was her man. He left that very afternoon and got a job at the lodge, a much better job than the one at SRI.

Since then, whenever I see Ephata he thanks me again and again for his good fortune. He is convinced that I saved his life.

I don't know if that's true, but I tell this story to express my profound astonishment at the moral principles that guide people like Ephata in the face of so much poverty and corruption. He could have handed the keys over to the robbers, and no one would have blamed him. He risked his life for an organization beneath contempt.

I'm just happy that his good name has served him so well. We expect everyone to be Samaritans, with no reward except a pat on the back, and then shake our heads whenever they fall from the straight and narrow. It is amazing that humankind has ever produced a single person like Ephata.

But Ephata is not unique. Africa is full of stories such as his, full of small, forgotten acts of kindness and courage.

SUNDAY, 10 NOVEMBER

Pam gets out of the car at Turner's Springs and guides Sarah across the rocky river bottom. Sarah turns sharply to the right, rolls slowly forward, then cuts back to the left. Pam directs her a little further to the left, then gives the all clear to straighten her wheels. Sarah pulls up on the opposite bank and Pam hops back inside.

The Loliondoes' signal is very strong. Sarah drives a quarter mile east along the Ngare Nanyuki and spots L23 sitting on the riverbank with another female, L51, who disappears into the gully as soon as we arrive. Inching up to the edge, we look down to see L51 with a tiny cub. Its eyes are still milky-blue, and it can barely walk on its stubby little legs; it is only a few days old. L51 wanders around the narrow riverbed, nibbling on long blades of grass. She delicately plucks off each blade and swallows it with great grimacing of mouth and tongue. Absorbed in her gardening, she disappears from view, and her cub vanishes underneath a low bush.

Leaving the lions hidden in their den, Sarah heads south and weaves her way through a dense stand of thornbushes. Brittle branches screech against the sides of the car, flinging broken twigs in through the windows, impaling the passengers in the back seat. Reaching the top of a low hill, Sarah drives in a tight circle, radio-tracking. Nothing.

We stop at the base of a tree-covered kopje, pull out a spare antenna, and scramble up the rocks, clapping loudly and watching for snakes and thorn branches in the tall grass. Near the top, we scale the sides of a large boulder, finding toeholds where thin slabs of granite have peeled off the sheer wall. At the summit is a clear view of three converging tributaries of the Ngare Nanyuki, each marked by its own line of trees. The countryside is rolling, rich, and tidy. The overall impression is lush and tropical, yet gently pastoral: Savanna-on-Thames.

Holding the antenna high above our heads, we pick up the faint signal of L81, a young male from the Loliondo pride. He is across the river, hidden somewhere far along the middle tributary of the Ngare Nanyuki. We spend half an hour looking for a shallow bank and a rocky bottom. Several spots look vaguely passable, but Christine and I are leaving for Arusha early tomorrow morning, and we don't want to risk getting stuck. Just as we are about to give up and go home, we stumble upon six Loliondo subadults draped along the riverbank. They are on their own, and they are a bit nervous about the car. Taking our time, we move in close enough to see who's who and spend the next hour updating their ID cards.

In the meantime, Christine has loaded up an ox heart with worm medicine. We have had no further success since MKP and MKQ. Hyenas have eaten our hearts without hesitation, but adult lions have all been too picky. Subadult lions scavenge throughout their nomadic phase, and have a more playful, adventurous spirit. Surely these six youngsters won't let us down.

By now the young lions have all moved to the opposite side of the river, about twenty-five yards away. Christine steps out and tosses the heart, but it slips from her gloved hand and sails only about five yards from the car. Everyone groans. The lions don't even look up.

After five minutes, I slowly open my door, crouch down, and creep toward the heart. I don't want the lions to panic and run out of range. But they are all out for the count; no one stirs as I slip back to the car. Concealed once more behind the open door, I heave the heart across the river.

My distance is good but my aim is poor, and the heart lands in a clump of tall grass about four or five yards upwind from the lions. They all sit up at the sound of the splat. One young male sniffs the air, stands, stretches, and heads toward the prize. But then he suddenly stops for some reason, changes direction, and walks over to the shade of a low stand of trees. One by one, the remaining subadults join him and settle down for the day, too far away for a second toss. We drive up and down the river for another half hour, but still can't find a crossing.

Foiled again, we call it a day and head for home, screeching through more brush and winding our way across country before finally reaching an old track. Halfway through the Loliondo kopjes, Sarah picks up the signal of the K2 pride, one of the many derivatives of the Loliondoes.

The entire pride consists of only two adult females and their two large cubs, and they are all watching something on the other side of a small stream. We stop the car about a hundred yards away as the collared female slinks off toward a small family of zebra. The cubs disappear under a bush. The second female, LLV, makes no effort to conceal herself. She lies on her front and watches her companion from afar.

The zebra stallion starts to whinny; there must be a rival male around somewhere. He snorts, and the whole family stares into the middle distance, too distracted by the complexities of their social life to notice the approaching danger. The collared female stalks to within twenty-five yards of her prey, then crouches down in the grass.

LLV stays put, still making no effort to hide or position herself.

Finally, the zebra rumble off with a loud clumping of hooves. The collared female charges along only a few yards behind, barely visible in the dust and grass. She keeps up for about a hundred yards, but never quite closes the gap. The zebra escape into the distance.

The chase over, the collared female stands panting for a few minutes, then walks back to the cubs. They greet her enthusiastically, rubbing up against her head and nipping at her heels. They all rejoin LLV, who seemed to find her companion's activities interesting but not compelling enough to join in herself.

Maybe LLV is feeling lazy today, or perhaps she is out of sorts. Or maybe she holds a very high regard for her companion's skills. This might have been an excellent opportunity for a single lion to succeed on her own.

I spent a long time thinking about cooperative hunting during the four-day follows, sitting in my Land Rover with pencil and paper doing my sums. Cooperative hunting was supposed to be ubiquitous in nature, but I had my doubts. I had seen too many performances like this one, and knew too many filmmakers who had spliced together solo hunts to make them look like group efforts.

I had asked one of my students in Minneapolis to photocopy all published articles on group hunting in any species that she could find. I kept them with me in a box in the Land Rover: reports on birds, bugs, spiders, carnivores, fish, and primates. They all hunted in groups. A few species were more successful when hunting together than when hunting alone, but most were not. At first, I couldn't see any pattern at all; some species seemed to cooperate whereas others did not. No one had a good explanation.

But then I started thinking harder about what a truly selfish animal should do during a group hunt, and it dawned on me that a great deal depends on the ability of a single hunter to catch its own prey.

If you are an expert big-game hunter, why should I ever help you? You have an excellent chance of catching supper by yourself, and the prey would be big enough to feed us both. Even if I'm as good a hunter as you, I couldn't possibly improve your chances by more than a few percentage points. Why should I lift a finger for such a small improvement in my own prospects for dinner? If I did, I'd have to get up, run somewhere, and maybe even get hurt. I'd rather let you do all the work, then join you at the dinner table. You'd be better off if I helped you out, but it would pay me to stay put.

However, if you are incompetent I can't rely on you alone. I have to help, even though I'm incompetent, too. I have to take the risk of the kick and the gouge, rather than just stand idly by and let you fail by yourself.

Reading through all the studies in my box in the car, it became clear to me that animals specializing on large prey seemed to cooperate only when they were incompetent. Skilled hunters were no more successful when they were together than when they were alone; they didn't seem to be working together as a team at all. Our own studies of the lions were consistent with this overall pattern, and Dave and I were able to show that groups of experts were indeed plagued by a breakdown in cooperation. Although female lions would almost always hunt together when a single hunter would be most likely to fail, they often held back when one hunter could get the job done.

Dave and I saw hundreds of hunts, and we discovered that females are most cooperative when hunting buffalo, but least cooperative when hunting warthog. Let a buffalo wander into view, and the entire pride would be up and at it. But should a warthog wander up, only one or two lions would stir themselves. Like LLV, the rest of the pride would sit patiently and wait for dinner to appear on the table.

This perspective also provided new insights into cooperative interactions between the two sexes. Female lions are famous for doing almost all the hunting in the pride; males are notorious chauvinists who watch the proceedings from afar. But males are big, and they lack the agile speed of the females. They have little to contribute to a group hunt of a wildebeest, warthog, or gazelle. Why should they get up from their easy chairs when there are so many more expert hunters to give chase? The females might appreciate some help, but they are going to succeed anyway.

Think of a doubles match where your partner is the top seed at the tournament, and you are a rank amateur. When the ball comes over the net, what are you going to do? Charge around and run yourself ragged, or hang back and force your talented partner to cover most of the court?

Now look at it from the pro's point of view. Are you going to force your incompetent teammate to pull his weight, or try to make the best of a bad job? In the presence of most males, female lions just have to grin and bear it.

Although the males live up to their lazy reputations most of the time, their greater size and strength gives them an advantage in hunting large, slow-moving prey. I once watched a group of females lead their husbands to a buffalo, then stand erect and literally point at the prey: There you are, dears; you can do something useful around the house for once. The males went dutifully forward, hopped on the buffalo's back, rode it like a bucking bronco, and finally pulled it down. Meanwhile, the females stood perfectly still, cheering the males on from a safe distance.

Lions would probably do better if they cooperated all the time, but individuals who do too much for the rest of their group are at an evolutionary disadvantage. He does best who does for himself, and leaves the most offspring to grow up behaving in precisely the same way. The expert hunters' dilemma is not that their companions won't assist them. Their problem is that they are wasting their skills in the presence of another expert; the two really should split up and hunt by themselves. Two of us chasing the same prey with absolute certainty of success will only gain a half-share each. If I can catch the same prey by myself at least fifty-one percent of the time, I would be better off working alone. Over time, one percent starts to add up.

It is tempting to take this principle too far, to consider so much interpersonal conflict in our own society as the outcome of arrogant self-regard. The Beatles effect: I'm so talented that I'd be better off going solo. But the parallels are difficult to avoid.

Figuring all this out provided some of my happiest hours in the Serengeti. Everything kept falling into place: species after species showed the same pattern, individuals cooperating only when they had to. The dusty articles sprang up at me from their box. Then the behavior of the lions fit the same pattern, a pattern no one had ever noticed before.

To feel that you have discovered something important about nature is intensely exciting. When the feeling hits, I can hardly sit still. My mind is on fire; the excitement sustains me over these rough, dusty roads, illuminates my cold winter days in Minneapolis.

Sitting here today, though, I find myself increasingly preoccupied by darker currents; currents that once again reveal similarities across species and that have been celebrated by poets and playwrights alike. The thrill of companions cooperating to annihilate their enemies, the overriding imperative of us against them, the great English victory at Agincourt:

> We few, we happy few, we band of brothers;
> For he today that sheds his blood with me
> Shall be my brother; be he ne'er so vile
> This day shall gentle his condition:
> And gentlemen in England, now a-bed
> Shall think themselves accurs'd they were not here,
> And hold their manhoods cheap whiles any speaks
> That fought with us upon Saint Crispin's Day.
>
> (*Henry V*)

For LLV and her companion, supper is merely a small domestic issue. They have more serious problems on their minds: they must protect their cubs from the constant danger of sexual menace, and they must survive the next confrontation with their neighbors. As two females against the world, they can only eke out a living in the cracks between larger prides. Last month they had three cubs, but they lost one last week. Perhaps it was killed by another pride, perhaps it couldn't keep up.

Last night they were near the Masai kopjes, today they are in the heart of the Loliondo pride range, and they won't be safe here for long. The mothers may not pull together for all the household chores, but they will protect each other until the bitter end. They may be doomed, but they have security in companionship. They are comrades in arms.

In Pam and Sarah's kitchen after lunch, the entire morning is defined by a moment of utter banality. I pick up a clean spoon, stir the coffee, and rinse it off in a basin. That precise sequence of events virtually defines the difference between our own world and the world we observe outside. In a truly selfish world, the sequence would always be reversed: wash the dirty spoon, stir the coffee, and then put the spoon down. That is, if you can imagine that there would ever be a kitchen in hell.

Out on the verandah, the birds stand on the table and encourage us to throw out more crumbs. A hyrax joins us in the shade, then flees in terror as Erin storms over from the Lion House. She has bad news. On her way back from Kenya, she saw two snared animals at Banagi Hill, a lame zebra with a wire dangling from its leg and a male lion with a snare around his waist. She thinks it might be possible to surround the zebra on foot and capture it by hand. She tells us where to find the lion and charges off to round up a team of park rangers. The rest of us prepare the darting equipment and drive north past the Seronera River, toward the base of Banagi.

To the west, the sky has built up over Lake Victoria. A flotilla of anvil clouds line the distant horizon. The tops are wispy-white, the bottoms heavy and black. The road ahead bypasses a bridge over the Banagi River that fell apart about ten years ago. A section collapsed under a car full of nuns.

"Always take the left fork and cross the river on this concrete drift. But remember, be very careful during the rains. It's hard to gauge the depth of these rivers when they rage, people are swept over every year. You may have to wait for days before you can get through."

As we drive along slowly, looking for the lion, we are suddenly aware of a constant tapping on all sides of the car. We are being besieged by tsetses. Dozens of large, dark flies dart in from every direction. Tsetses thrive in these hot, humid woodlands and they carry trypanosomiasis, or sleeping sickness. In their way, tsetses provide vital protection to the park—wild animals are unaffected by sleeping sickness, but the disease is fatal to domestic stock. Tsetses make vast portions of the Serengeti uninhabitable to farmers and pastoralists.

Erin had said the snared lion would be right up here on the left side of the road, about two kilometers from the river. Yes, there he is, just a few feet off the road. He's a fully grown male, and he's consorting with a female.

The female stands up and walks about ten yards along the edge of the road. The male follows close behind. The female lies down on her front and growls a continuous low rumble. The male straddles her and mounts. He ejaculates as soon as he penetrates her—the king of beasts has a hair trigger. As he squats, throbbing, he gives off an eerie yowl and chews the back of her neck. Once finished, he steps to one side. The female slowly rolls over onto her back, one foreleg stretched straight out, the other paw on her chest.

Although they look like any normal mating pair, he is damaged goods. A wire snare has cinched his waist, just in front of his hind legs. In pulling himself free, he yanked the wire tighter and tighter until it dug deeply into his flesh, crushing his abdomen. He is as barrel-chested as any male lion, but now he has a twelve-inch waist. It is hard to believe that blood can circulate past the constriction; I don't see how he could ever defecate again.

Looking through binoculars, we can see that he is no longer carrying the snare. Having pinched his waist, the wire has fallen off. There is nothing we can do for him now; the damage is already done. But where did he pick up the snare? We are about five miles outside our study area, and we have never seen him before. Has poaching moved this far inside the park? Or was he snared in the far west or north? Perhaps he has started roaming far and wide, determined to devote his last ounce of energy to sex.

Killing an animal with a gun or an arrow at least involves some degree of choice, but snares are indiscriminate. No poacher intends to catch a lion or a zebra in his snares; they are set out for smaller animals such as impala. Larger animals get caught just the same and ruin themselves in breaking free. The Serengeti holds much food for the impoverished people around the park, but any harvesting has to be directed. Focused, efficient, and humane—not like this.

Pam starts the engine, and the noise inspires the couple to another performance. Lions mate every half hour throughout the four days that the female remains receptive; the male lion's reputation stems from his stamina, not his self-control.

The yowling, throbbing, and rolling is an exercise in utter futility. Even if she conceives, there is no way that he will be able to live long enough to protect his cubs from the next set of males.

We turn back toward Seronera and meet Erin on the road by Banagi. The zebra had moved off by the time she arrived. She leads the way home, and we stay far enough behind to avoid her trailing plume of dust. The western storms have dissipated, the sun sinks behind a spent wall of cloud.

After dinner, the air still redolent with crushed rosemary, we sit around the coffee table in Pam and Sarah's living room and try to come up with a better method of worming the lions. The room is bathed in soft, cool light from two solar-powered fluorescent bulbs. A large aquarium sits in one corner, filled with fish from Lake Tanganyika; its aerator bubbles silently. The Debussy cello sonata drifts up from the tape recorder. Outside, a yellow-winged bat hangs from the crossbeam above the verandah. It darts at the windows, catching bugs attracted to our lights.

We know that the lions will eat ox hearts if they taste them, but they don't seem to trust a lump of meat that has fallen from the sky. George Schaller provisioned several lions during his study; he gave one male a seventy-five-pound chunk of buffalo meat to see how much a lion could eat. The old park wardens used to shoot wildebeest to feed starving cubs.

"But those are all familiar foods."

"Lions don't often encounter isolated beef hearts."

"Maybe if we could make the heart move somehow. They do sometimes play with objects that dangle from the car."

"That's it. We can tie the heart to the end of a string, throw it out the window, and play cat and mouse. We'll have time to go out early tomorrow morning before leaving for Arusha. If it works, we can ask Barbie to send some more hearts down with Rob when he gets back."

"By the way, whose turn is it to wash up?"

I walk back to the Lion House in the dark of midnight, my flashlight casting a shaky white circle on the narrow path ahead. Low in the sky, the half-moon glows weakly. The night is quiet. No lions have roared around here for days, and even the crickets seem to have run out of steam. Way off to the north and the west, somewhere beyond Banagi, a long, yellow line blazes brightly on the black horizon, a grass fire raging out of control. The arsonists were probably cattle rustlers. Masai warriors often sneak across the park at night, raid the tribes on the western side, and flee back into the wilderness, setting fire to the countryside behind them to cover their tracks.

The god Lengai gave all the world's cattle to the Masai. Other cattle-tending tribes are merely temporary guardians of the Masai's rightful property, and the Masai can reclaim their stock whenever they wish. In retaliation, the Wa-ikoma, Wakuria, and Sukuma mount their own spirited raids against the Masai, usually taking more animals than they lost.

The Serengeti serves as a no-man's-land in the middle, a theological battlefield. Large stretches of woodlands have been lost to the fires that burn through the park. Once upon a time, these disputes held a certain primeval charm, but now both sides are exceedingly well armed. They no longer rob their opponents with spears and arrows but with semiautomatic weapons.

And each new victim adds a death that must be avenged. The flames of hatred are fanned with ever greater intensity. The inferno burns higher and higher against a blank wall of darkness.

MONDAY, 11 NOVEMBER

"Do we have any more gloves?"

"Thread the twine through the heart and wrap the outside once or twice."

"Drop the spool onto this screwdriver, and we can let out the line."

"Okay, let's go fishing."

All four of the Loliondo males are here today, sitting in an uneasy semicircle around L23. L23 gets up and crosses the river; she is being courted by MS14. Her mate for the day follows at her heels. The other three males move over to sniff the spot where she has been lying. They lift their heads into the air, upper lips raised, teeth bared, tongues sticking out. The facial expression is called a "Flehman face," and its contortions enhance their sense of smell.

The couple have settled by a tree on the other side of the river. The other males join them and collapse under a nearby bush. We drive to the nearest crossing and double back on the opposite bank.

"Approach them very slowly. We can't afford to disturb them; we haven't got much time."

No one has moved by the time we arrive. The other three males are lying in a small thicket twenty yards from the consorting pair.

"Head over to the right and circle around. Stay as far from the lions as you can . . . This is close enough. Toss it over by that bush.

"Nuts! The grass is taller than it looks. Try reeling it in."

The grass twitches as the bait stutters across the ground. One of the males watches intently.

Tug, wait, tug again. The lion loses interest.

We yank the heart six feet toward the car. All three males sit up, but they watch along the length of the line. The twine has risen from the grass like a low clothesline. Whatever this may look like to the lions, it doesn't remind them of food. One of the males finds us too irritating, stands up, and slowly ambles down into the gully, gone for the day.

"Reel it in and try again. Let's go over to the consort pair."

We squeeze back through the brush, out into a clearing, then drive in a wide loop back to L23 and her suitor.

We toss the heart again. Yellow eyes gaze blankly at us instead of at the bait. Curiosity certainly won't kill these cats.

Twitch the line. Tug, yank. Heave.

Blank.

"Come on, lions, where's your spirit of adventure?"

"Drive by and trawl for them—drag the heart along behind."

Nothing.

We double back, try again. Still nothing.

"Okay, let's drag it past the other two males. Wind your way through those bushes."

We drive right by, and neither lion shows a glimmer of interest. We stop the car. The lions stand up, stretch, and join their buddy in the gully. The tattered heart, covered with dirt, has caught on the trunk of a small bush and torn loose from the twine.

Back at the Lion House, the luggage is piled into our silver Suzuki, a flimsy little car that should probably be buried in an unmarked grave. It is only four years old, but it has kept Rob stranded at the Seronera garage for most of the past year. Starting the car is like turning on a sewing machine; the sound of the tiny engine lacks the authority of a Land Rover or Land Cruiser. Erin swears the Suzuki is mechanically sound right now and that only the bodywork is falling to bits. We have to trust her judgment about the mechanics, but she is obviously right about the bodywork. We can see the ground through the rusty floor.

Christine and I wave good-bye to Erin, wish Pam and Sarah good luck in their first few weeks of independence, and drive off in the bright, cloudless morning. A perpetual cloud of dust engulfs us as the car rocks and bounces along the road. By the time we reach the Seronera River, we are already filthier than if we had driven a full day in the Land Rover. The bounces and bumps are merciless; the corrugated surface has been gouged out by an endless parade of wheels much larger than our own. The stiff little suspension was never designed for roads like these—no wonder the car has shaken itself to bits. If we try to go faster than about forty, a continuous deafening, crashing sound tumbles down from the roof.

We clatter past the Simba kopjes, then on toward Naabi Hill. Here is the spot where one of my Tanzanian students rolled his own tiny Suzuki. According to witnesses, he had a blowout while going at least seventy. His car flipped over three times, but he escaped unhurt. He showed up at the Lion House late that night, powdered white with dust, his clothes shredded, his eyes as wide as a ghost's.

We had given him a car for his research project in the northern Serengeti, but he had been running a transport business instead of watching lions. He fabricated his data while carrying goods around the park.

Although we had been warned about him by the SRI staff, he had always seemed so enthusiastic, so knowledgeable about day-to-day life in the field. But when I finally sat down and looked over his data from the preceding year, it was obvious that he had forged most of his observations. He had never seen a lion by itself, only groups, and all the groups were happy families of just one male and at least one female. I went down to the Seronera garage, checked the log book and found that he always managed to observe as many lions during the long weeks when his car was being repaired as when he was on the road.

I confronted him with the evidence and he confessed to everything. When I asked which of his sightings were real, he could no longer remember. He had spun such a complex web of deception that he had gotten lost. Nothing could be salvaged from the little work that he had actually done. We felt betrayed, of course, but we were most disturbed by his indifference to being exposed. He felt no shame; he was only upset at losing his precious car.

A harsh, peppery smell suddenly rises up from the back of the Suzuki. Oh, hell! Christine's bottle of formalin has started to leak. Get this stuff out of here! It can pickle your lungs and cloud your corneas. We tie the bottle onto the roof rack, and in the process find the source of the crashing noise. A heavy tow chain has been coiled loosely under a spare tire.

Chains, tires, and chemicals rearranged, we carry on past Olduvai, past the turnoff to Laitoli, site of hominid footprints laid down in volcanic ash three million years ago. We bounce, judder, and shake past Masai families who have arrived in anticipation of the rains, bringing their cattle to what they hope will soon be greener pastures. We crawl up and over Ngorongoro Crater stopping every half hour to readjust the formalin bottle, which keeps tipping on its side and dripping on my shoulder. The combination of dust and formalin has turned my hands into bear paws; they feel like sandpaper gloves. Layer after layer of dust blankets the inside of the car: white followed by red, now white again as we enter Mto-wa-Mbu.

We make a short diversion to the headquarters of Manyara National Park to look for the new chief park warden, Ettha Lohay. Ettha is the first and only female CPW in Tanzania, and she has been a dear friend for many years. She was a junior warden at Gombe in the mid-'70s and one of the true heroes during Gombe's darkest hour.

Ettha is the mother of three children. Her first pregnancy was somewhat ill-timed, and she was hastily transferred to an office job in Arusha. But she is such an exceptional warden that she was soon brought back into the field, and has since served as warden throughout the country. Ettha was born in this region, and Manyara is her favorite park.

We exchange huge hugs, and delight in seeing each other again. Holding hands in that warm African manner, we ask after each other's families and children. When I ask after her oldest daughter, Ettha mentions the woman for whom her daughter is named, a petite American woman who was with us at Gombe. When I reply that the woman is doing just fine, Ettha breaks out laughing, reminding me that I had previously told her of the American's marriage—to a small man.

She says, giggling uncontrollably, "And you said 'She finally found someone who is the same caliber.'" We laugh, and Ettha is almost bent double, stamping the ground, hooting "The same caliber!" Her junior staff peer at us from around the corner, but there is no self-consciousness to spoil the moment.

Bouncing, rattling, and jolting along the washboard road in the late afternoon, we acquire a final coating of bright red dust. Crawling gingerly onto the asphalt, the Suzuki is on home turf, and we buzz smoothly along past granite hills and cinder cones tinted purple and orange in the dying light. The approaching sentinel, Mount Meru, soars high in the sky but is barely visible in the haze of dusk.

Darkness falls as we enter the outskirts of Arusha. The few streetlights illuminate dim circles on the sidewalks, and our filthy headlights barely cast a beam. Pedestrians wander half-seen in the shadows; the familiar daytime landmarks lie hidden in the gloom. We somehow find our way back to the same hotel, the same phone, and a simply unbelievable shower. Innumerable muddy streams pour down onto the gleaming porcelain tiles.

ARUSHA / TUESDAY, 12 NOVEMBER

After breakfast, we visit the head office of Tanzania Game Trackers, located in the middle of a coffee plantation just outside Arusha. We park our puny little Suzuki at the end of a long line of luxury Land Cruisers that take Western hunters on safari.

Adam Hill welcomes us to his office and shows us a large sack of snares that has just been delivered from their hunting zone in the Maswa Game Reserve. People in the local villages are rewarded for each snare they collect.

"They found these in just the last couple of weeks, and we don't even operate in the worst areas. The western end of the Maswa is simply crawling with poachers. They must really be hammering it over there."

The sack is overflowing with hundreds of handwoven loops, each made of plastic rope. Attached to most of the snares is a small wooden wishbone, a trigger that holds down a branch until an animal steps on it. Before now, I'd seen only wire snares; the poachers obtain the wire by burning old tires. Wire is much stronger than plastic, but at least it's hard to find. This new rope is woven from the countless plastic sacks that transport grain and fertilizer to every small village in the country.

"Last year we paid out twelve thousand dollars for snares, bows, spears—we even got a couple of rifles." Adam shakes his head, "The contents of that bag alone would kill ten times as many animals as our clients take in a year. We also ask our clients to pay a surcharge of twenty percent on top of their license fees. The money goes to a local village for a development project. The village closest to us decided to buy a maize mill this year."

In Africa, hunting is still part of the social fabric. But the human population around the Serengeti has grown exponentially. People are crammed all along the western border, and they are hungry. Limited numbers of animals could certainly be cropped each year from the Serengeti, but unregulated slaughter would decimate the last great herds of wildlife on earth. The park is not an island safely adrift in the Coral Sea. The Maswa Game Reserve provides an important buffer along the southwestern boundary.

The future of the Serengeti depends on the security of the surrounding areas, and on the goodwill of the people who live there. In exchange for the licensed harvest of a small number of animals by tourist hunters, at least one large area is being protected from the mad, random slaughter, and the local people are encouraged to protect their own resources for the future.

Having said all this, though, I still find trophy hunting exceedingly distasteful. I was brought up to eat what you kill, and I simply cannot imagine why anyone would put a lion head on his wall. But lion hunting is very big business in this very poor country. Allowing wealthy hunters to shoot a few male lions each year brings in large sums of money and doesn't threaten the future of the species.

Since I can't stop lion hunting, I at least want to see some good come from it. If these animals are going to die, we ought to try to learn something from their deaths. Tanzanian Game Trackers has permission to shoot about a dozen lions per year, and we want the hunters to collect tissue samples from their victims. Christine could obtain parasites from the lions' intestines, geneticists could extract information from their skin and muscle tissues, their bones could tell us something about their nutritional state. Adam is willing to help, and we leave him with a dozen small boxes filled with bottles, preservative, and instructions.

On our way to the airport, we pass hand-painted signs for the Zion Apostolic Mission and the Holy Ghost Fathers, then turn right at the sign for "Lake Duluti Club, Camping Paradise." We wind through yet another coffee plantation, then grind up a wide, flat-topped hill along a narrow track. The Suzuki seems somewhat reluctant and coughs once or twice, but having cleared its system, it soldiers on up to the top. The hill turns out to be the wall of a volcano, a flooded caldera nearly half a mile across.

The view from the top is exceptional, even for Tanzania. To the west, Lake Duluti forms a brilliant green circle surrounded by lush, steep walls. To the north stands Mount Meru with its crumbling peak and massive green flanks. But the crowning glory is Mount Kilimanjaro, floating in the haze to the east. Four or five posh houses dot the crater rim, and we drive up to an old British colonial-style house with a low roof, stone walls, square windows, and outsized shutters.

"*Hodi.*"

Henry Fosbrooke steps outside from his study and invites us to join him for coffee. In the late '50s, Henry became the first conservator of the Ngorongoro Conservation Area, and he was instrumental in defining its boundaries. The colonial government had originally intended to include the Crater in the Serengeti National Park, but this plan would have uprooted the Masai from their sacred lands.

Proud, independent, and indifferent to luxury, the Masai resisted the seductions of a Western lifestyle. Preferring to follow their nomadic, cattle-herding traditions, they did not hunt wild game and practiced no agriculture. To many, the Masai were an integral part of the savanna ecosystems. So the NCA became a unique experiment, a "multi-use land area" where the indigenous Masai have been allowed to live among the protected wildlife.

Henry came to East Africa in the '30s to join the colonial administration, and he ardently admired the Masai. To protect their cattle from the migrating herds, he built a fence along the boundary between the Serengeti and the conservation area, but the wildebeest simply knocked it down. He planned to build a Masai school on the floor of the Ngorongoro Crater, and he even left his British family to marry his Masai housekeeper.

Today most conservationists feel that there are too many Masai in the NCA. It was set aside for descendants of the families that lived there in 1959, but this rule has proved impossible to enforce. As animal husbandry improved, the Masai herds became far larger than anyone expected, and many Masai now rely on goats as well as cattle. Almost everyone worries about overgrazing by the Masai stock, but not Henry. He thinks that the problem is too many wildebeest.

Henry has become a precious resource. He has continuously documented the changing vegetation around Ngorongoro, and he was the first historian of the Crater lions. Henry is in his eighties now, and I try to see him whenever I pass through. But no sooner have we completed all our greetings and introductions than the phone rings. It is an Egyptian with a dream. He wants the Masai to replace their cattle and goats with camels.

Camels fare better in arid country than do cattle, and they are less destructive to the vegetation. The Egyptian doesn't say how he plans to convince the Masai to abandon their ancient traditions, but he engages Henry's attention for what seems like hours. By the time Henry finally hangs up, it is time for us to leave.

We bounce back down to the paved road and continue on toward the airport. Before the days of colonialism, the Masai pursued their cattle-rustling campaign across this dusty plain with no real restraints from neighboring tribes. On the lower slopes of Mount Kilimanjaro, children of the Chagga tribe would serve as lookouts and warn their parents of impending Masai raids. The Chagga dug an extensive system of tunnels with cavities large enough to conceal whole herds of cattle. Huddled like termites in secret chambers at the base of the volcano, they would remain underground for days on end.

The stakes were high. Besides cattle, the Masai also stole women.

The parking lot at the Kilimanjaro International Airport is almost empty. Escaping the heat, we enter the terminal and wait for our aircraft to arrive from Dar es Salaam. The large, shiny building echoes like a steel-plated barn; the only flights today are on Air Tanzania and Ethiopian Airlines. The snack bar serves warm beer and soda, and the peanuts are wrapped in unmarked plastic bags. Many of the acoustical tiles are missing from the high ceiling, most of the fluorescent lights have burnt out, the hardwood floor is splintered and worn.

The airport was built in the early '70s in anticipation of a boom in international tourism before the final imposition of Tanzania's socialist policies. By the time the airport was completed, the Tanzanian government had become suspicious of tourism—wealthy Westerners might corrupt the peasants and the socialist state.

As the years went by, the contrast between the privately owned Kenyan lodges and the state-run Tanzanian hotels became so dramatic that few Westerners felt two weeks in Tanzania qualified as a holiday. More and more tourist traffic came through Nairobi rather than Arusha. The big tour operators promised to pamper their clients in Kenya and then whisk them down to Tanzania to see the glories of the Serengeti, Ngorongoro

Crater, and Lake Manyara. But they always paid the Tanzanians in local currency rather than in the Western currencies so valuable for foreign trade.

The Tanzanians resented the Kenyans for gaining all the foreign exchange from tourism when Tanzania was so much more spectacular. So the Tanzanian government closed the border to build up their state-run tourist corporation, and the Kenyans eventually retaliated by closing their side of the border. For seven years, only long-term residents were allowed across, and even they had to obtain police permits from Dar es Salaam and Nairobi.

As fiscal reality hit hard in the late '80s, the Tanzanians decided to encourage privately owned tourist operations. They finally reopened the border, but their tourist industry had not yet become competitive. Local companies lacked the capital to purchase new vehicles, and years of isolation had insulated them from the realities of market forces. Today, the Kenyan tour operators continue to dominate the Tanzanian tourist industry, and while Nairobi receives four or five European flights each day, European airlines no longer fly to Kilimanjaro.

Air Tanzania Corporation is also known as Any Time Canceled, so we are uncertain whether we will actually be going anywhere today. But, surprisingly, our plane arrives on time. It is a small blue-and-yellow jetliner with a giraffe painted on the tail fin. We find our seats on board, and the African stewardess walks down the aisle, handing out copies of the national newspaper, the Tanzanian *Daily Nation.* National news focuses on corruption and the logistics of distributing agricultural supplies throughout the country in time for the rains. The country's only political party, *Chama cha Mapinduzi* (The Revolutionary Party), reaffirms its support for socialism while the country's president, Ali Hassan Mwinyi, continues to press for economic reform.

On page three we discover that the outside world still exists, after all: the Soviet Union is close to dissolution, Yugoslavia is on the brink of civil war, Somalia has collapsed into anarchy. Without the sideshow of ideology, history has finally resumed—the history of a fractured, disordered planet.

The plane is still a thousand feet above Dar es Salaam, but the cabin is already hot and stuffy. On the ground below, small patches of forest and scrub are surrounded by hundreds of small sandy fields. Tree-lined streams meander through the barren patchwork like snakes. Glistening flooded swamps stand poised to send countless swarms of mosquitoes into the low-lying shantytowns that encircle the city center.

By the time we taxi along the Dar runway, I am drenched with sweat. Outside, a coastal sensibility asserts itself: move very slowly and don't expect to get anything done.

Part II

Dark Star

Dark star crashes

Pouring its light into ashes.

Reason tatters

The forces tear loose from the axis.

Searchlight casting

For faults in the clouds of delusion.

—The Grateful Dead

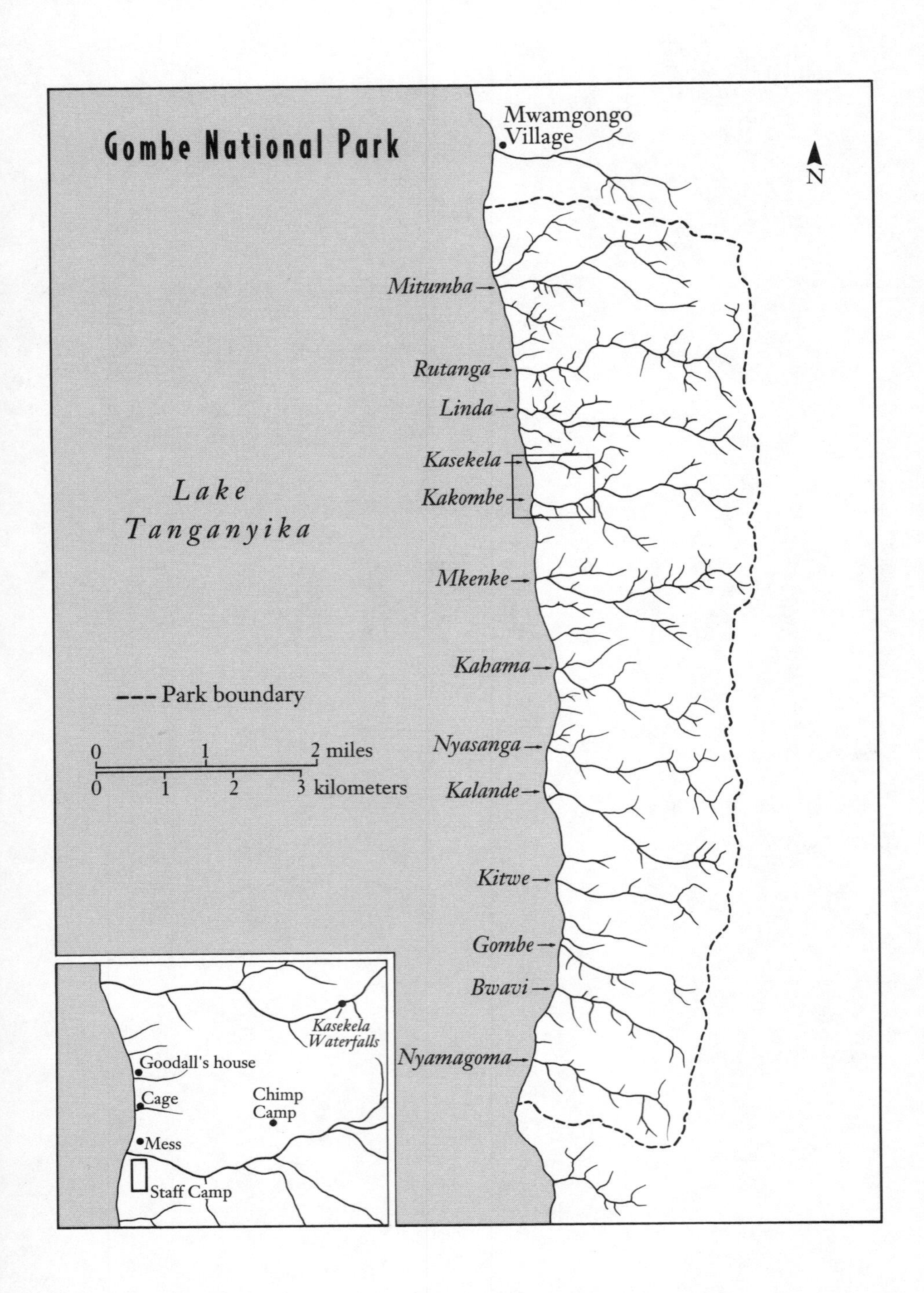
Gombe National Park
Mwamgongo Village
N
Mitumba
Rutanga
Linda
Kasekela
Kakombe
Lake Tanganyika
Mkenke
Kahama
Park boundary
0
1
2 miles
0
1
2
3 kilometers
Nyasanga
Kalande
Kitwe
Gombe
Bwavi
Nyamagoma
Kasekela Waterfalls
Goodall's house
Cage
Chimp Camp
Mess
Staff Camp

DAR ES SALAAM / WEDNESDAY, 13 NOVEMBER 1991

I wake to the sound of a worn-out air conditioner rattling in my hotel window. The curtains are drawn, and the sun is already up. Feeling slightly leaden from the change in altitude, the heavy damp air, I stagger down the linoleum-covered stairs to join Christine at breakfast. Just as I sit down at the table, the skies open and pour with rain. The morning is stiflingly hot, the rain falls like sweat.

Hepatitis and amebic dysentery flourish in these coastal kitchens, so I decide to order a Coke after surveying the cracks in my teacup. We are in the restaurant of the Mawenzi Hotel, refuge for many of the frugal expatriates who travel to Dar on business. Dar es Salaam is Tanzania's capital and its principal port; everyone spends time here eventually. The Mawenzi is only two blocks from the center of town. The aging air conditioners don't always cool the rooms, but they stir the air and mask the noise of the streets.

Peter Jones joins us to pick up the keys to the Suzuki. He lives in Arusha and has come to Dar to purchase vehicles for his safari business. He and his wife first came to Tanzania as anthropologists but changed professions to be able to stay here forever. They caught *le mal d'Afrique*, the obsession with Africa, the horror of returning to the dull, temperate world where summer comes and goes. They first tried alternative agriculture, harvesting the native plants of the Masai steppe. Then they turned to tourism, and did so well that Peter was able to start his own company. He will be flying back to Arusha this afternoon, and he can store our Suzuki somewhere safer than the airport parking lot.

Breakfast over, Christine and I step out into the muggy morning and skirt around the muddy puddles as we search for empty cardboard boxes to take to Jane Goodall's house. We dart into each shop and office along the street: Accurate Supplies, Tanganyika Traders, Morogoro Stores. No one would throw away a good box in this country, but no one seems to have any to spare, either. Finally, Christine manages to extract one from the Goethe Institute, and I locate another at a large pharmacy by the Askari Monument.

The monument commemorates the African foot soldiers who fought in World War I. Tanganyika was a German colony until 1918, and the country has retained much of its *Deutsche Ostafrika* infrastructure. Although the European population here is far smaller than in Nairobi, Dar is surprisingly cosmopolitan. Many of the original German settlers endured Tanganyika's annexation by the British Empire, and substantial numbers of Norwegian and Swedish aid workers came out during Scandanavia's infatuation with Tanzanian socialism.

Although Tanzania is a member of the British Commonwealth, it remained staunchly nonaligned throughout the Cold War. China built a railway; Canada and East Germany sent advisers and engineers. But no one country has left much of a mark. By straddling the ideological divide, Tanzania has received more foreign aid per capita than any other country in Africa, and, as most people here are eager to tell you, they have the least to show for it.

Boxes in hand, we walk to a nearby taxi rank. After negotiating the fare, we hop into a Peugeot that is at least twenty years old. The only surviving window handle sits in the ashtray, but the seats have been reupholstered and the dashboard is lined with plush red velvet. The African driver starts off carefully and eases out into heavy traffic, coaxing his precious car along.

The city buildings are almost all concrete. Most are grey, some are a bleached yellow; the constant damp air corrodes the ornate, Eastern facades with dark patches of mildew. But the teeming, crowded streets are alive with color. Most of the women have wrapped themselves in brightly printed cotton cloth, and many carry bundles on their heads. At least half of the men wear a *balagashia*, the brocaded cap of Islam.

We pass the city jail where white-shirted prisoners sit crowded behind iron bars that extend all the way down from the ceiling. They dangle their legs outside and yell taunts at passersby. A fierce stench billows out into the street.

Stopping briefly for fuel, the driver points to the attendant who is cranking the gas pump by hand. He tells us that the city government is bankrupt, the municipal power supply only operates at night. Back on the road again, we roll slowly past Government House, the old colonial hospital, the shady lanes of diplomatic houses. The United States embassy bristles with antennas and video monitors. The French embassy bristles, too, but somehow it manages to bristle with style.

We pass over a narrow estuary, stinking at low tide, then carry on for a couple of miles past impromptu roadside markets and rows of small kiosks, all shiny and new, selling cashews, rice, and Kenyan dry goods. The driver turns cautiously onto a road that must have been used for testing mortar shells. He had winced when I told him our destination, and he calculated our fare not so much on distance as on the depth of the potholes. We drive past Julius Nyerere's house and turn right.

Nyerere was the first president of Tanzania and the architect of its social and economic policies. Unlike most African leaders of his generation, Nyerere was neither overthrown nor rich when he left office. He is known to everyone as *Mwalimu* (teacher). Now living in modest retirement, he is revered for his personal honesty and integrity even though his well-meaning policies were widely despised by the end of his reign. Many Tanzanians say, "He didn't want anyone to get rich, so he made sure that everyone stayed poor." Nyerere may have had a flawed vision of human nature, but as a moral figure he was as admirable a founding father as any nation could ever hope for.

We bounce down a narrow sandy road almost to the beach, and here I am again, no longer the captain of my own team, but a minor player for someone else.

We have arrived at Jane Goodall's headquarters in Africa, the house built by her second husband, Derek Bryceson. Derek was a close friend of Nyerere and one of the few white members in the Tanzanian parliament until his death ten years ago. The house is right on the beach; the Oyster Bay Yacht Club is just across the sound. On a still day, you can hear the muffled explosions of peasant fishermen dynamiting the offshore reefs.

Jane Goodall has flown to America to lecture and to raise funds, but her American assistant Chris greets us in the guest wing. Chris works for both the World Bank and for Jane. She hands us a letter from Anthony Collins, Jane's chief of staff at Gombe and the principal collaborator on our own primate grant. Anthony has asked us to collect his research clearance and a few jars of peanut butter.

We walk to the main house in the shade of the palm trees and casuarina pines, enter through the vine-choked verandah, and slowly climb the stairs to a wide room cluttered with shelves, steel trunks, and filing cabinets. The air is dense, the floor damp. The room reeks of fungus and rot. Every available surface is covered with tall stacks of files dated somewhere between 1960 and 1990. The stacks are labeled as "Attendance," "B-record," "Swahili Notes," and "Monthly Reports."

The files may look tattered and moldy, and their labels may seem obscure, but they contain treasures of inestimable value. Here before us stands the sum of all the knowledge that Jane, her students, and staff have gathered every day for over thirty years. This dank room stands as a testimony not only to Jane's own perseverance and self-discipline, but also to the teamwork that she has inspired. Page after musty page was dragged up and down steep mountainsides by people who were willing to risk life and limb, to suffer serious illness, and to devote years of their lives to the chimps and baboons.

Jane and her students have used only a portion of these observations in their publications. Her team has observed the animals for so long that the achievement has become an unmanageable mountain. Then, too, Jane has greatly expanded her efforts in the past few years. No longer just that extraordinary figure who watches chimpanzees in the wilds of Africa, she devotes herself to the protection of chimps throughout the world and campaigns for reform in biomedical research. She travels throughout Africa, Europe, and North America each year and has little opportunity to focus on her ever-accumulating accomplishments in Dar es Salaam.

Anne and I were students at Gombe when the system of data collection was perfected, but we were only able to keep a small hand in the primate studies after we took over the Serengeti lion study. We have recently received a new grant from NSF to oversee a complete analysis of these long-term data and to initiate new studies at Gombe. First, though, I have to move the mountain to Minnesota.

I throw open the upstairs doors and windows, hand Christine a notebook, and make an inventory. Cockroaches, termites, and silverfish slither out from the middle of the stacks, and the older paper crumbles around the edges. We escape the stuffy room every hour or so, going down to the verandah to cool off in the sea breeze. Chris joins us to commiserate about the heat and the state of the country, and asks how long we will be in town.

"Until the day after tomorrow."

"Who is going to ship it all?"

"Don't know yet."

In the late afternoon, we complete the inventory just as a yellow minibus drives up. It is a local businessman, an Asian who has been helping Jane with a construction project. He is about to go back into the city and can give us both a lift. He says he has a cousin who can pick me up first thing tomorrow morning and help me box, transport, and ship the data before lunchtime.

"What's his name, Aladdin? Does he rub a lamp and make a wish?"

"No, his name is Rasul. You pay him in cash."

THURSDAY, 14 NOVEMBER

At 9:00 A.M., cousin Rasul arrives at the Mawenzi Hotel in a Toyota pickup truck. We make our way down India Street, toward Morogoro Road, to the heart of the business district, past Bigdeal, Ltd., Bojak Store, Jal Ram Traders. Many of the interiors are dark—the power is already off for the day—but the goods shine through the gloom of each open doorway: hardware, clothes, medicine, tinned food, and farm supplies.

The fruits of imported enterprise lie glittering in the rubble of a lost civilization.

The city streets are as pockmarked as the road to Jane's house. Rasul fumes at each crater. "The streets haven't been repaired in over twenty years. The government sets aside money for roadwork each year, but the road commissioners put it all in their pockets. Cars, buses, lorries—all are being torn to pieces!"

Rasul parks by a busy corner. Nearby, half a dozen men stand beside various large stacks of cardboard. This is the box market. The street merchants invite us to inspect flattened, bundled boxes that brought pharmaceuticals, chemicals, and machinery to Africa.

"How much?"

"Two dollars each."

Rasul objects, "That is too much; you should only pay half that."

A proud old man shows us his fine German boxes—precision cardboard, no doubt. One dollar fifty apiece. Rasul makes a face, but I haven't got all day to save a few bucks. Besides, the old man is a Green entrepreneur.

The sky is dotted with cotton clouds, the air crystal clear, the sunlight sharp and hot. We wedge our way between pedestrians and potholes, past mosques, shops, and more shops. Vegetable stands line the pavement of one street; another is heaped with burlap bags of charcoal. None of the traffic lights work. A large orange bus stalls in the middle of an intersection, the slogan on its rear window reads "Really No Problem."

Rasul wants to know how much it costs to live in Minnesota. How do rents compare with salaries? He has family in Canada, but it is too expensive. He has family in England, too, but there are no jobs. Things are changing in Tanzania. "There are plenty of opportunities here these days. It is such a mess, but somebody has to fix it. There is a lot of work. I have a large family here, and we have new business every day. If the government goes back to the old foolishness, I may go to Uganda. I have family in Kampala."

After surviving the final few miles to Jane's house, Rasul and his African assistant follow me upstairs to the data room and help refold the boxes.

We stack file after file of intimate biography into each carton. The social intricacies of famous personalities: who mated with whom, what they ate, where they went, the battles they waged. The chimp equivalent of *War and Peace*, the baboon equivalent of the Domesday Book, but unedited, scattered. Raw material.

Files from the '60s, then the '70s, fill one box after another. But then somewhere in the middle of 1975, a repressed anxiety forces its way to the front of my mind. I have to sit down.

Better to lay the ghost before I go back. Turn to the month of May. Leaf through to the fifteenth, sixteenth, seventeenth, eighteenth.

Here it is: 19 May, just another day like all the rest.

But then a chill. Written at the top of the form is the observer's name: "Smith." It was Steve's day at the feeding station. He wrote down the name of each chimp that came through camp, the direction each came from, what each did while it was there. But no mention of what happened to Steve that very same night.

Turn the page; a gap until 21 May. One day off, then back to normal—except that from then on the observers' names are all African.

But there is no time for reminiscence today, and I don't want to start getting cold feet about returning to Gombe, either. We quickly stack the next sixteen years of chimp files into another few boxes, give the baboon data their own container, and the job is done. The pickup truck strains under the heavy load as we bounce back to the middle of town. At the shipping office, Rasul presents me with a bill for the morning's work and says good-bye.

The shipping agents couldn't be more helpful. The boxes weigh seven hundred pounds. Lufthansa has an afternoon flight with direct connections to Minneapolis, and everything should arrive late tonight. The air freight charges must be paid in foreign exchange. A credit card will be fine. Here's your receipt, thank you very much. The bill is less than half of what I expected.

Hey, is that all there is to it? This was supposed to be the big headache that would only be half-finished by the time I left the country. It always used to take weeks to achieve anything in Tanzania, but all this pent-up energy has suddenly been released. The economy has been opened, and people have been granted a powerful incentive to cooperate—with those who can afford it, of course.

It is noon, we leave for Gombe early tomorrow morning, and I've got only one free hour to go shopping. None of the shops have anything I need. One after the other, the Asian shopkeepers send me up and down Samora Avenue: "There you will find it." But the shops are filled with cheap toys, Western-style clothes, bootlegged cassette tapes, and musty souvenirs. The display window of a government-owned building is empty now, but during the dreary days of socialism it was piled high with all the electronic gadgets forbidden by economic restrictions. There were stereos, radios, and cameras, the only ones in town, kept safely behind heavy steel bars. Nestled among the goods was a large sign that read, "Nothing in this shop is for sale."

After lunch, we head out to the University of Dar es Salaam. Christine has had a successful morning, too, and she has arranged to meet a parasitologist at the zoology department this afternoon. The university is several miles out of town, and our driver has agreed to bring us back this evening. He turns on the cassette deck, and a sudden jolt of African pop music tumbles out of his lavish loudspeakers.

"Do you like the music?"

"Sure, play it as loud as you like."

It is Kanda Bongo Man, a band from Zaire. The shuffling rhythms, jangly guitars, and punchy brass sections push us through the endless traffic jams in the middle of town. We finally pick up speed and cruise onto the highway. To the left, an open-air market offers nothing but secondhand clothes donated by Western charities. On the opposite side of the road, craftsmen sit in the scant shade. Tire carvers convert treads to sandals. Tinsmiths manufacture trunks, charcoal stoves, and oil lamps. Children play with elaborate toy cars made from string and wire.

Grinding up a steep hill, we pass through the gate to the university campus. A friend of mine used to live in that house over there. Thieves broke his door down one evening and stole his stove, refrigerator, and a large pile of paper—he had brought out his own stack of research data from England.

He lost years of careful labor. In utter despair, he went for a walk beyond the campus grounds the next day and strolled through the woods to a small shantytown. At the market, he bought some peanuts wrapped in a paper cone. Unfolding the cone, he recognized the markings on the paper. It was one of his data sheets. He eventually recovered all of his data, but he had to buy a lot of peanuts.

Jim Parkin is an old Africa hand. Originally from Edinburgh, he now heads the parasitology lab at the university, and he has studied the worms of monkeys, people, livestock, and wildlife. He knows how to get his samples, and he is full of practical advice. "You have been going about it the wrong way. Give the lion some male fern extract—it induces vigorous diarrhea in about ten minutes. Most of the adult worms are spewed out with everything else. It's quick and effective, but I don't know if anyone still makes it.

"For primates, you can probably get by with a strong chocolate laxative. It may not work as fast as the fern extract but it will get the job done. Failing that, however, put your sample on some washed charcoal. The charcoal absorbs the toxins in the feces and lets the eggs hatch. Just stir in your sample and keep it warm—easy enough to do in this climate—and it'll be crawling with larvae in a week or so."

Our driver selects a tape of Tanzanian music for the ride back to town. Each song follows the same basic pattern: the singer makes his point for three or four verses, then stands aside to let the drums and guitars catch fire for the next ten minutes—manic, hypnotic, flashing with neon streaks of colored lightning. Most of the lyrics reflect the usual preoccupations with love and family, but the rest are public service announcements about AIDS, or keeping the streets clean, or remembering to wash your hands after using the toilet.

The late afternoon is cloudy and the heat penetrates everything like an oven. We reach one of the big pharmacies at the Askari Monument ten minutes before closing time, only to find that they have nothing on Parkin's list except chocolate laxative. The pharmacist, a petite Asian woman, asks, "What is the age of the child?"

"Say what?"

"I need to know the child's age to calculate the dosage."

"Oh, no," Christine explains, "this is for monkeys."

The pharmacist giggles uncontrollably, wasting precious seconds before finally pouring the tablets into a small, unmarked pillbox.

Then we make the measured yet frantic walk to a surgical supplies shop over on India Street, careful not to walk too fast and risk suffocation in the viscous air. We leap in through the door just as the *askari* starts to lock up for the evening.

"Litmus paper? Male fern extract?"

"No, we don't have any."

"Formalin?"

"Maybe down the street."

Everyone has now closed for the evening, and we knock vigorously on door after door—Janoowalla Chemists, Modern Pharmacy Limited, Mansoor Daya. The shopkeepers all let us in, but to no avail. A city filled with pharmacies, but each one carries the same few items: Malaraquin, Hedex, Dispro, and Panadol. If you don't have malaria or a cold, you're out of luck.

Defeated, we retreat to the She Self-Service Supermarket, its blue-yellow fluorescent lights flooding the darkened sidewalks. The air-conditioned aisles are filled with extravagant imports from Asia, Europe, and the Middle East. Shelves sag under the weight of tinned Christmas puddings from England, Peruvian anchovies, diet soda from Pakistan, bottled water from Saudi Arabia, and Swiss breakfast cereal—but no peanut butter for Anthony.

In the steamy night, the blackened streets are transformed. Only mad dogs, businessmen, and tourists walk around Dar in the midday sun, but at night the Africans jam the pavement. Makeshift vegetable stalls appear next to small open fires roasting bananas and ears of corn. Old friends greet each other, and children stretch their legs.

I am usually nervous walking these city streets at night. Nighttime is the worst time for street robberies. Anne and I were once robbed at knifepoint in Dar, and a friend was slashed just outside the American embassy. But tonight we are surrounded by families strolling in the open air. A bustling, cordial community has miraculously sprung forth.

We enter the air-conditioned lobby of the Agip Hotel for our last meal in civilization. The Agip has the best restaurant in town, and their telephone is usually in order. I need to ask Barbie Allen to relay a message to Anne in Minneapolis. While the receptionist tries to call Nairobi, we wait in the bar surrounded by sun-scorched Europeans; an African pianist plays outdated Western standards.

Christine is feeling a bit more optimistic tonight. The time in the Serengeti was not good for her self-confidence, and the conflict with Erin has not yet been resolved. But Jim Parkin was helpful and enthusiastic about her project. Gombe will be much easier for her: the baboons eat a high-fiber diet, they poop every few hours, and they will eat chocolate laxatives like candy. And Gombe is crawling with parasites, one of several reasons why I am so ambivalent about going back. Primate pathogens thrive in the hot, damp forest, and most are infectious to humans. I was constantly ill during the long rainy months at Gombe. And I mean ill, more ill than I like to remember.

The receptionist waves at us from the front desk; my call has gone through. The line hisses and crackles. Standing in the wood-paneled lobby, surrounded by wealthy customers, I start shouting at the top of my voice.

"Barbie, hello! Everything's going well, we're off to Gombe tomorrow. Can I trouble you with a few things? First, could you please fax Anne and tell her that I shipped all the Gombe data this morning. It should arrive in Minneapolis late tonight . . . Yeah, I'm amazed, too.

"Second, could you possibly call around Nairobi and see if you can find some male fern extract. That's M-A-L-E F-E-R-N extract. It's a fast-acting laxative made from male ferns, presumably. It makes you shit so fast that the worms get thrown out with it."

Here I am, standing in the snootiest restaurant in Dar es Salaam, screaming over a lousy telephone line about shitting lions. This kind of behavior is a common failing of field biologists in public. We forget how we sound to the rest of the world.

I remember one particularly entertaining reunion with Gombe friends in an English pub. Spirits were high and voices were raised during an animated discussion of the baboons' sex lives, and astonished onlookers had no idea we were talking about monkeys. "Heathcliff and Juliet were copulating on the beach when Gangster came over and chased Juliet out from under him. She leaped out so fast that Heathcliff fell on his butt! Ha!"

"Did you ever see Moses play with himself? Jane once told me about how he used to sit on her windowsill while she cooked breakfast each morning: 'He would masturbate just a few feet away from my stove. Then he would sit there and eat his ejaculate. It put me right off my scrambled eggs.'"

FRIDAY, 15 NOVEMBER

Our taxi to the airport has seen better days. All of the gauges and indicators are broken; exhaust fumes pour in at our feet. The engine sputters and stalls, but the driver takes it all in stride. He hops out of the car while the traffic backs up behind us. Angry motorists start honking their horns as he asks some pedestrians to give us a push. Half a dozen members of the "self-employed informal sector," as they are officially known, fall in behind. We are soon rolling along in silence until the engine starts up with a groan and a belch.

It is only 7:00 A.M. Our plane doesn't depart for another two hours, so we have plenty of time. Our driver starts to talk to us; he is a Zanzibari with two wives and seven children. He tells us how much he likes foreigners.

"I am very happy to see people bring business to this country. Tanzanians want to be able to work and to get paid for it, to buy and sell things wherever they want."

"How old are you?"

"I don't know. Fifty, sixty. I like fifty-five. Say I'm fifty-five."

We pass the Shri Swaminarayan Temple, its entrance guarded by plaster lions with golden manes. Next comes the Cultural Center of the Islamic Republic of Iran, then a large billboard that reads: "TOP CLUB CIGARETTES for men whose decisions are final."

Parked in the loading zone by the departure desk, our driver helps pile the luggage onto a trolley. The airport was built by the French about ten years ago. The high, fan-vaulted roof blocks the sun and the rain, the elegant glass walls reach halfway up from the ground, the rest is open air. Birds nest in the roof beams above the departure lounge.

At the check-in counters there are long queues for flights to Nairobi, Lusaka, and Paris, but nothing for Kigoma, the gateway to Gombe.

"Has the plane to Kigoma been delayed?"

"No, it has already left."

"What?"

"The plane has gone. You had better check with the Air Tanzania office."

"Excuse me, we have tickets for today's flight to Kigoma."

"It departed three hours early."

"Three hours early! How were we supposed to know that?"

A nervous giggle, "They announced it on the radio last night."

"This is not funny. How the hell do we get to Kigoma?"

"The next flight is on Tuesday. Otherwise take the train."

"The train is always fully booked, and anyway it takes three days to get to Kigoma. People are expecting us today. I don't believe this." Utterly defeated for the moment, I am tempted to throw it all in, go to Zanzibar for a few days, and just sit on the beach. But while casting about for some miraculous solution I remember all those budding capitalists we have been seeing everywhere.

We are six hundred miles from Kigoma. If we can charter a plane this morning, we might still reach Kigoma before Anthony takes the boat back to Gombe this afternoon. Otherwise we will be caught in the domino theory of African travel, with each delay causing us to miss each subsequent connection.

"Is there a charter company around here somewhere?"

"Go to Terminal Two."

We find another taxi and drive to the airport built by the British during colonial days. I leave Christine in the car and run inside the crumbling labyrinth. Once more, the hard problem is easily solved. It only takes half an hour to find a pilot who is flying to western Tanzania this morning. He will be happy to double the length of his flight and take us to Kigoma, at a price that will absorb every penny I managed to save on the air freight charges. We depart in two hours.

Now we need to send a message to Anthony. An airport official leads me through a series of grimy, abandoned rooms, then into an office occupied by six people who seem to have nothing to do. "Where do you need to call?"

"Kigoma."

"Just one moment." A secretary in a red nylon sweater turns an ancient telephone dial with the eraser of her pencil. Everyone gathers around to watch. No response. She dials again. Then the supervisor walks in. Everyone hastily returns to their desk, and the secretary sits on the number. The supervisor glowers at me and asks one of his subordinates what I am doing in here.

"He is waiting for his pilot."

The supervisor sits uneasily at his desk for a few minutes, pretending to work, then stalks out again. The phone number reappears from underneath the secretary. "It is against the rules for us to make private calls."

She dials again. Still no luck. We will be taking off soon. Perhaps Chris could relay a message to Anthony before we land. The secretary tries calling Jane Goodall's house, but no luck there either.

She tries Kigoma again, but the supervisor pops his head through the door, and the game is up. I am instructed to leave immediately, and the staff prepare themselves for an angry tirade.

Christine and I move our luggage to the tiny departure lounge, pay the airport taxes, and fill in the forms. The room reeks of public toilets that have never known running water. The pilot leads us out to the tarmac and helps us climb aboard his small twin-engine aircraft, a six-seater.

Airborne at last, we soar above the coconut palms and banana trees, the swamps and sprawling cityscape. Towering thunderstorms burst along the diminishing coastline. It is raining heavily over on Zanzibar, an immense grey storm squats above the green and blue. We finally escape the low coastal heat and leave the ocean behind.

An hour later we are over Dodoma, the arid city that is scheduled someday to become Tanzania's new capital. The rains haven't arrived here yet; the earth is still pinkish-brown. The surrounding countryside has been denuded by overgrazing; cattle trails envelop each inedible bush like cobwebs. Dodoma is also home to Tanzania's only wineries, where they make Dodoma Sweet Red. There used to be a large billboard in the

Dodoma train station: "Dodoma Red. All the subtlety of a charging rhinoceros."

After Dodoma we follow the rail line west. The railway, built by the Germans before World War I, follows the Arab slave-trading route to Kigoma. We can still make out the widely spaced line of mango trees that provided shade every few hours along the long march. Though no one seems to talk about it much any more, the Arabs plundered East Africa for slaves for more than a thousand years. African slaves drained the swamps of Iraq in the ninth century. Ending the slave trade was one of the principal justifications for European colonization.

The pilot starts his descent into Tabora for refueling, flying low over a crazy quilt of half-acre fields deeply furrowed by wooden plows. Each small plot has its own round termite mound, raised like a nipple on the nurturant earth. Circling above the airstrip, we see a large crowd gathered beside the boarding ramp. As we taxi off the runway, a brass band starts to play, and the crowd waves banners and flags. Our pilot tells us that they are expecting the Tanzanian prime minister. When we step from the plane, the crowd realizes its mistake, the band loses steam, and everyone wanders back home.

The air is fresh and clear; Tabora sits at the western limit of the central plateau. From here on, the rivers drain west toward Lake Tanganyika through wild country that was once the center of the ivory trade, an empire ruled by Mirambo, the "Napoleon of Black Africa" as Stanley called him. At Tabora the railway divides: one line goes north to Lake Victoria, the other continues west to Kigoma. Transit passengers to and from Kigoma always spend a day here awaiting their connecting train.

There are great train trips to be taken in Africa. The train from Nairobi to Mombasa, for example, is a remarkable experience. But the train to Kigoma is an experience to survive rather than enjoy. The rolling stock is dilapidated and overcrowded. Thieves wait along the line and poke hooked sticks through the open windows to yank out the passengers' belongings. The toilets may have been cleaned once upon a time, but it is hard to be sure. Many African rail travelers spend the entire journey drinking beer, which puts a lot of pressure on the facilities.

Our train trips between Kigoma and Dar were always packed with local color. Stinking bags of dried fish lined each corridor, chickens and goats wandered in and out of the compartments, children's pee would drip down from the upper bunk onto the person below. African passengers who didn't know we spoke Swahili would blurt out angry anti-Western, anti-white comments.

But one particular trip still stands foremost in my mind.

Anne was accompanying a naive young woman from Stanford University (I'll call her Theresa) who had just arrived in Tanzania for a six-month stint as an undergraduate assistant to Jane. Anne had already worked at Gombe for several years and knew her way around. She acted as Theresa's guide and chaperone; they boarded the train together in Dar and shared a first-class compartment.

Theresa was very excited and talkative, and thought that it was just so nice to chat with all the friendly African men in the bar—utterly oblivious of the implications of her behavior. A group of drunks tried to resume interactions in the middle of the night by knocking loudly on Anne and Theresa's compartment door. Theresa gaily got out of bed in her night-gown and started to unlock the door. Anne stopped her in the nick of time and told the eager crowd to try their luck elsewhere.

This nags at me because Theresa, an exceptionally outgoing, flaky young woman, went on a holiday to Zaire in 1975, just a few weeks before the crisis at Gombe. She had the time of her convivial life, presumably letting everyone within earshot know all about this fantastic little research station in the middle of nowhere, full of young, undefended foreigners.

Anne and I were in Nairobi when we first heard the news. A friend called our hotel at 2:00 A.M. and told us that a group of armed men had come across Lake Tanganyika from Zaire in the middle of the night. They had abducted four students from Gombe.

At first we thought they had taken only women. We had frequently heard that gangs of Zairian men would roam the lake and steal women from other villages. We were convinced that Theresa had made such an impression on the far shore that she had inspired some drunken throng to raid Gombe.

It wasn't until the following morning that we discovered the kidnappers had taken Steve as well. When the captors' real motives finally became clear, we actually felt somewhat relieved. At least there would be some chance that we might see our friends again.

The pilot calls ahead to the control tower in Kigoma, announcing our imminent arrival and finally relaying our message to Anthony. West of Tabora the land is flooded and wild. The Malagarasi River snakes and loops below in a fantastic tangle of oxbows and finger lakes dotted with floating papyrus islands. But the swamplands of the Malagarasi do not stretch as endlessly as I remember from years past. Within twenty minutes, we have reached the other side. The wilderness is being eaten away by roads and villages. Scattered fields are lined with cassava and corn, the irregular patchwork is streaked with angry red wounds of erosion.

Flying in a wide arc, the pilot slips between two massive rain clouds towering ferociously above our flimsy aircraft. Safely through, we catch our first view of Lake Tanganyika, where Stanley met Livingstone, that Burton mistook for the source of the Nile, home of the *African Queen.* Formed by a narrow rift, Lake Tanganyika is one of the oldest, deepest lakes in the world.

On either side, great green escarpments rise thousands of feet above the lake's placid surface. Ahead and to our right stand the scalloped peaks of Gombe. A line of low mountains thrusts straight out from the water on the opposite side, the mountains of Zaire. I try at first to keep my gaze on the world around us. But the pilot is approaching the red dirt airstrip in the same pattern that Anne and I took in May 1975.

We came back to Kigoma to collect our data and to reunite with all the Gombe refugees. We flew in with a CIA agent from the American embassy in Nairobi. He was a pompous little man in his starched shirt, silk tie, and dark trousers. He had spent the last two years as personal attaché to Henry Kissinger in the Middle East, and had arrived in Nairobi only three weeks ago. "Things seem pretty straightforward out here; the Kenyan security forces are well run. The country reminds me of Lebanon."

"That may be true in Nairobi, but there is no security down here. You have to rely on your wits and the goodwill of the local people. Have you ever read *Heart of Darkness*?"

"Who wrote it?"

"You've come to Africa and you've never read *Heart of Darkness*? It's by Joseph Conrad. It's about what happens to people when they are separated from civilization—how the human spirit becomes corrupted in the absence of moral guidelines. You see those mountains over there? That's Zaire; it used to be the Belgian Congo. The Congo was Conrad's symbol of a land completely cut off from modern civilization. We are landing in full view of the heart of darkness. Kigoma is about as far from Nairobi as you can get."

In those days there was only one small building at the end of the dirt airstrip. The fire truck was a tattered Land Rover painted bright red. When we taxied up to the rickety little terminal, there were no cars waiting for us. Even the fire truck had broken down. Our CIA man wasn't worried, though. He said he had sent a Telex to the chief of police, asking to be met as soon as he landed.

But no one came. Half an hour went by and we still hadn't convinced Mr. CIA that no one would ever come. His message to the police chief had probably been lost or, more likely, ignored.

"Why not call a taxi?"

"No, I'm sure that the police will send a car. They must be tied up with the kidnapping."

"Why not call and ask?"

"Good idea." He turned to the lone airport attendant who had been napping behind the desk and asked him slowly in special English: "What is the number for the local police department? Poe-lease dee-pawrt-mint."

The attendant sat frozen with embarrassment and fear until I translated the request into Swahili. He immediately replied, "Three," and started cranking an ancient windup phone.

It was about then that the stuffing started to seep out of Mr. CIA's shirt.

Today, though, our communications have been more effective, and a Land Rover is here to meet us. The Kigoma air terminal has been expanded, too. They have a proper control tower here now, even a real fire truck and a baggage scale. Ten or fifteen people are loitering around the base of the tower. One of them wears a maroon T-shirt that says, "Looking through the past to see the future."

In the Land Rover we hold on for dear life as the African driver races around huge puddles in the middle of the red mud track. The wattle-and-daub houses are the same color as the road; some are roofed with thatch, others with corrugated iron. Papaya trees poke above the courtyards and dense, round mango trees line the street. Small children wave gleefully at the car and shout "*Wazungu! Wazungu!* (White people! White people!)"

In a tiny village Christine buys charcoal for her parasite samples. We reach the tarmac and speed into Kigoma, down the familiar main street, past the market and all the old shops. Everything looks the same as always. At the end of the street, right on the waterfront, stands the German-built train station. The ancient white paint is stained almost as red as the muddy road. The clock tower in the roundabout says that it is eight minutes past midnight. Lord knows how many years it has been eight past midnight in Kigoma.

Mr. CIA had finally talked to the police but they weren't interested in him, so he asked the airport attendant to call a taxi. We came into Kigoma and turned off at the police station. There on the verandah sat the American consul to Zanzibar, overweight, lounging in shorts and sandals, sipping a beer. Our man launched himself out of the taxi, charged up to the consul, and stood at attention. "So-and-so reporting for duty, sir."

With a look of utter disbelief, the consul stared at him for a second, belched, and asked: "What the hell are you?"

Dismissing Mr. CIA, the consul turned to us and told us that the rest of the Gombe students were about to leave on the train to Dar. The research center had been evacuated. There would be no more foreign researchers at Gombe. Given our own experiences, Anne and I were concerned that the students would find the journey highly traumatic so soon after the kidnapping. Four friends had been captured for god only knew what reason. All our assumptions about the safety of our private little world had been way off base. The train would provide no comfort to anyone.

We reached the station just a minute or so before the train pulled out. Our friends were jammed into a single compartment. We stood on the platform and talked through the open window with hardly enough time to find out what had happened. Helen and some of the others were crying, and the vultures had found a feast. Half a dozen American reporters encircled us, trying to photograph the students in their distress. I lunged at one of them and nearly knocked his camera out of his hands.

"Can't you see they're upset? All you want is to prey on their feelings."

The cameraman fired right back. "Look, who's taking advantage of who? The police won't let us go out there. Nobody will tell us what happened. This is a big story. And you're going to be flying out tomorrow on the plane that we chartered from Nairobi."

The train started to move. We ran alongside, grasping the hands of our departing friends as the train gathered speed. And the photographers got their tearful pictures. They followed us for a while after the train had left, but we refused to talk. Mr. CIA had told us to remain silent.

Today, the Land Rover follows a muddy track along the edge of the harbor. We pull into the offices of Jane's agents, yet more cousins of Rasul, and they tell us that Anthony is still in town. No one has seen him to give him our message, but the Gombe boat is still here, and several of the park's staff are already on the beach waiting to go back.

We pass the long afternoon under the cool shade of a mango tree. After four hours, Anthony arrives laden with food and supplies, astonished to see us. When he discovered that we had missed our flight, he assumed we wouldn't show up until next Tuesday.

Anthony loads everyone into the twenty-foot wooden boat and paddles away from the shallows. The outboard engine stalls and sputters in the dying sunlight before finally settling to a constant roar. We plow out into Kigoma harbor, churn around the rocky point, and follow the coastline north into the sunset.

Thirty miles away, the mountainous silhouette of Zaire broods above the reddened lake. To our right, the Tanzanian coastline is still warm in the final glimmer of daylight. The fields and trees soon mingle in the gloom, and the jagged escarpment blots out the starry horizon. The middle of the lake is spangled with shining beacons skating on the inky surface. Fishermen have paddled out in their dugout canoes, seeking *dagaa*, the freshwater sardine. Their kerosene lamps attract the fish to their nets, but in another few days they will be outshone by the waxing moon.

Cooking fires throw a flickering orange glow against the mud-brick houses that line the shore. We cross the invisible park boundary, and the landscape is lit only by the cold light of the moon; the dark forested valleys hang suspended between dim mountain slopes.

In the murk, even after all these years, I can recognize each valley: Kahama, Mkenke, now Kakombe. We draw closer to land, chugging past square concrete buildings shrouded by trees and shrubs, past the sheet-metal house where I used to live. Anthony cuts the engine, and we are engulfed in sudden silence. The boat grinds on the gravel beach in front of Jane's verandah. Warm lake water laps at our legs as we drag the hull onto the silver shore.

I walk a few yards away and stare into the empty night, nervous of the shadows in the dark. Even after all these years.

GOMBE / SATURDAY, 16 NOVEMBER

We trudge along the beach in the dully, misty morning. The lake is quiet beneath a grey sky. The stones underfoot have been polished by surf, bleached white by sunnier days. We parallel a sheer cliff crisscrossed by creepers and vines. Above and behind us, forest and clouds block any view of the rift escarpment.

We reach a row of brown buildings nestled under a grove of tall trees. The windows contain no glass, only coarse wire mesh to discourage the grasping hands of the chimps and baboons. We climb a set of steps near the entrance to the old dining hall—we used to call it the Mess. Embedded in the topmost step, barely visible beneath a layer of wet gravel and sand, lies a mosaic of a chimp, David Greybeard. I don't know why, but I spend the next few minutes sweeping the step with my hands, clearing away the accumulated detritus, and bringing the portrait back into view.

There were always six to eighteen students at Gombe. We would assemble at the Mess each evening to eat and read and talk and drink. We ate together at a long table, with place mats and napkin rings. We ate a lot, always hungry after following the animals up and down steep slopes for hours a day. When Jane was at Gombe, she always sat at the head of the table. Over in the far corner was a large aquarium filled with fish caught by her son, Grub.

The Mess was where we recounted our adventures. Chimp news: "The males went down to Kahama today, hunting colobus monkeys. Figan and

Gigi were with them, but Figan couldn't decide whether to stay with Gigi or join the hunt. He had the hots, but he wanted an hour with the boys." Baboon news: "Hamlet was near B troop in Plum-tree Thicket, but then Jonah came tearing out of nowhere and chased him all the way across Kakombe Stream. This afternoon, I found him back home in C troop, grooming with his mother."

The Mess was also a place for fun and games—poetry recitals inspired by the shocking promiscuity of adolescent female chimps, silly skits about the sexual frustration of adolescent males. There were organized games and songwriting contests, farewell parties, birthdays, Christmas celebrations, romantic liaisons, good times and bad.

Now the Mess serves as a shabby tourist hostel for the national park. It has been divided by flimsy plywood walls into a half dozen cubicles, each with a rotting cloth door and cheap metal bed frame. You can smell the moldy mattresses from here, as ripe as a good Roquefort. Most tourists come just for the day; they are not encouraged to spend the night.

Continuing on our way, we walk along the row of offices and greet the chief park warden. We explain what we will be doing in the next ten days: Christine will be collecting fecal samples from the baboons; I will be setting up the chimp scale and completing the groundwork for a new map of the park.

Then we find Apollinaire in the baboon office. Apolly is the headman of the baboon project. He taught me how to recognize the baboons when I first came here in 1972, and he also taught me Swahili. I have the very deepest respect and admiration for this man. Everything he does is impeccable. He can be quite contemplative and philosophical—at times, almost poetic—but he has an objective, no-nonsense attitude about his work. Apolly's most impressive talent, though, is his ability to follow the baboons through the forest without ever letting a hair get out of place. Whenever the baboons go romping down a steep ravine through thorny vines, I always end up with muddy trousers, torn shirt, scratched face, and mangled shoes. Apolly comes out immaculate, ready to dine with royalty.

Preliminaries completed, Apolly, Christine, and I walk together over the narrow footbridge above Kakombe Stream. Two women down by the streambed are filling buckets with drinking water. Our progress to the staff camp is sporadic, for I stop frequently to talk to people from the old days, and we take time to catch up on each other's news. To the staff I'm still Bwana Hi-guy.

I used to pass through here several times a day in my callow youth, going to and from the baboons. Rather than repeat the long ritual of "*Jambo*" and "*Habari*" and more "*Habari*" twice a day with everyone I met, I contaminated local custom by insisting on a quick exchange of "Hi, guy!"

The first questions on everyone's mind today: How is my wife? How many children? They are very polite when I tell them two, but their thoughts are obvious. Not too bad for an *mzungu*, but what's wrong with you people?

The village is much larger than it used to be; there must be at least a hundred people living here now. Once home only to the African research staff, the park rangers were moved here as a security measure after the kidnapping, and more houses were built or moved down the hill. Each student used to have his or her own one-room hut in the forest between the lakeshore and the chimp feeding station. Now our old huts sit between the narrow blocks of staff quarters.

At the far end of the village we walk through a wet field of grass into the midst of a baboon troop. Apolly tells us that it is D troop. The baboons are feeding in a series of scrubby young fig trees, looking a bit sorry for themselves in the steady morning drizzle.

I point to a young female and Apolly recites her name and lineage: she is the granddaughter of an animal I knew well during my study, and she is in estrus. Her hindquarters are red and grotesquely swollen, and she is being guarded by an adult male. She stops at the base of a tree a few feet away from us. The male mounts her, clasping the backs of her legs with his feet and the small of her back with his hands. His elbows locked straight and his shoulders hunched, he enters her with a remote, distant expression. He might as well be irrigating the moon.

After a few slow, deep thrusts, several juvenile baboons run up screaming. The female lurches out from under her partner, moaning and grunting as if launched by a spasm of automatic ecstasy. The male raises his eyebrows, exposes his canine teeth, and charges a few feet toward the youngsters. They all scramble away. The female climbs up a small tree and calmly starts eating strawberry figs.

A baboon troop is somewhat similar to a lion pride. Females are usually recruited into their mother's group, and males move from one troop to another. But unlike lions, male baboons move around as individuals. Each Gombe male leaves his troop of birth and then moves by himself into a new troop to find the females of his dreams. Male baboons form temporary alliances with other males, but these relationships are nothing like the lifelong bonds between male lions.

Baboons are more variable than lions when it comes to sex. Some male baboons settle into a particular troop for life. Other males flit from group to group, searching ceaselessly for the next bursting female bottom. Some males seem to fall in love with certain females; others mate with whomever comes into heat. Family men and gigolos, all in the same small society. Like an endless fraternity party with no one watching, young males unregulated by parental guidelines or any other form of authority.

In the beginning of the Gombe baboon study there were two troops, Beach and Camp. Both troops divided shortly thereafter. Beach troop split into B and A troops. Half of Camp troop stayed in this valley, the other half migrated to parts unknown. A, B, and C troops coexisted reasonably well for the next ten to fifteen years, but B troop continued to grow and finally split into B and D troops a few years ago. Now B troop contains far fewer females than D, and Apolly is worried that B will be exterminated by its rivals.

Apolly has reason to worry. After all, the original chimp community divided in two, and the larger group subsequently annihilated the smaller. And, too, Apolly is from Burundi. He is a Hutu. Step down to the beach for a moment and look toward the north. The clouds have lifted. Burundi juts out from the eastern shore of the lake just fifteen miles away, a former Belgian colony that only makes the news when the Tutsi tribe tries to annihilate the Hutu. After each new onslaught, thousands of Hutu trudge past Gombe on their way to refugee camps near Kigoma.

D troop has headed up the cliff, and we follow close behind, grasping wet vines and tree trunks, pulling ourselves up the slippery slope. The gentle whooshing of the lake recedes to a mere whisper. The occasional low grunts of the baboons stand out from the buzzings of the crickets and cicadas. At our feet, ghostly white mushrooms, wispy fractal ferns, and virgin blades of grass glow against the black leaf litter like celestial bodies bursting with living energy. All sense of scale is lost; the tiny panorama seems endless.

A baboon leaps onto a tree limb, sending a shower of water down on our heads. The troop carries on up the hill, and I point to different animals, asking Apolly for an identification. He recites each name and its mother. The mother's name usually evokes a flash of familial recognition: "Why, he/she looks just like her."

We recognize these animals just as we do people, by their faces, and just like people, there are family resemblances: she has her mother's jowls, I remember her grandmother's baggy eyes, his mother's long nose. But each animal also has its own individual features: he has short hair and a square head, that one has a heavy pelage like a lion's mane. Pinched face, flat face, long muzzle, weak chin, square chin, buck teeth, wide nostrils, bottle snout.

It is easy to tell the baboons apart once you learn how to recognize their faces. But their distinguishing characteristics aren't something you can write down with precision like the lions' whisker spots, and they don't always show up in photographs. The knowledge is stored in the minds of a corps of observers, the resident biographers and historians of Gombe.

Being from Burundi, Apolly is fluent in French. He can communicate directly with Christine and orient her while I tend to other chores. I clamber back toward the cliff top. There should be a path down to the beach just over here somewhere. The bushes form a wall of leaves, the bright young greens stand out from the dark mature foliage. Every now and then, the green is punctuated by startlingly red berries that defy anyone to eat them. A swallowtail butterfly flits about, its wings studded with turquoise checks. As I slither beneath a large bush to make my way down the cliff, my hand brushes a stone marker.

More chills: this is Ruth's grave.

November 17th

Dear Anne,

I am writing from chimp camp, looking out the window at the dial of the new hundred-kilogram scale. If any chimps come through in the next few hours, I'll be able to tell you how much they weigh. I spent a long time looking for a decent scale in Nairobi, but couldn't find one that was waterproof. I am trying to establish a new procedure whereby the followers only set it up when they want to weigh someone. The old scale sat in the rain for ten years and was completely corroded.

It is *very* wet here right now, but the forest is extremely beautiful with all the fresh green leaves on the trees, the proliferating vines and ferns. Anthony is frustrated by all the paper work he's had to do lately, but he has just found a new administrator who will relieve him of many of his worst duties. Anthony is clearly essential to the staff's morale. There is still a remarkable sense of loyalty here, and the spirit and dedication are truly outstanding.

Today's observer on the attendance record is named Msafiri. Patti has just arrived with her family, and—oops, we hung the scale too close to the tree trunk, and she grabbed the banana without climbing the rope. But now Msafiri is climbing up to move the chain and hang the scale between two trees so widely spaced that Patti won't be able to reach the bucket from either tree.

Here she comes again, and, damn! She's climbed along the cross-chain and yanked the banana out with her foot. Damn chimps. Msafiri's verdict: Patti is a bit nervous of the contraption, but another chimp might climb the rope okay. Evered and Beethoven came through earlier this morning; most of the rest are in Kahama today.

I still have one more banana in case somebody else passes through. Once I get all this sorted out, I'll start hiking around with my GPS and look for palm trees. I saw Markus in Nairobi, and he has agreed to redo the aerial photography of the park. Anthony thinks June would be the best month, and I'll try to finalize all the details in the next few weeks.

Msafiri and his afternoon replacement, Warimu, are now dangling from the scale, weighing themselves. The mood is good. We need another chimp. If this setup doesn't work, I'll have to go to Kigoma to get some building material.

Craig Stanford and Charlotte Uhlenbroek are both here. Charlotte leaves for Europe tomorrow and will post this letter. Craig is here for another month or so. He is finding fewer, smaller troops of red colobus monkeys than when Clutton-Brock was here in the late '60s—he suspects they've declined because the chimps eat so many of them.

Now Sandy has arrived. She's nervous too, but the final banana is in place. Here she comes and, hell! She's pulled the same trick as Patti. Time to move the chain.

I've just climbed on the roof and found a good bolt to which I can anchor the chain. This will make a span wide enough to prevent anyone from climbing along and bypassing the rope. Now I need to go get more bananas.

Back from lunch. No chimps have been through since Sandy. I'm listening to a bunch of noisy tour guides from Kigoma who've just arrived with a tourist. The tourist is edgy, wondering if he'll see a chimp. Will any nearby chimps be attracted to camp by all the chattering? What will they make of the tourist's lavender trousers?

At lunch, Anthony distributed a four-month supply of birth control pills to one of the chimp followers. Jannette brought out one-hundred-and-twenty-months' worth. But how to distribute them? Hilali has three wives and twelve children, Yssa four and nineteen, Bofu five and twenty. There aren't enough pills to go around.

Yesterday afternoon I went with Apolly and Christine to the Dell, where we found a lot of *Biomphalaria* snails. All of the Kakombe valley baboons have schistosomiasis, and the drinking water still comes from the stream. We've brought some water purification tablets, but it's hard to remember to use them when you get so hot and thirsty. Gombe still seems as unhealthy as ever. Anthony has had another bout of malaria since he came back out; Charlotte and Craig got jiggers from the sand fleas up at Mitumba. The wet ground around here looks ideal for all the other nasties.

Thunder. As the rain starts to splash down, Warimu carries a bench outside, climbs up, and takes down the scale. In it comes, still dry. This system may work out after all.

The rain also pulls in the tourist, who turns out to be Dr. Hans Rosling, associate professor in the International Child Health Unit at Uppsala University. He conducts research on iodine deficiency in the people living over behind the rift. Improper preparation of cassava root impedes absorption of iodine, and the poor soil around here is very low in iodine anyway. Over five percent of the women have goiter, and one to two percent of the babies are born with mental retardation. He has brought out 250,000 iodine tablets to solve the short-term problem.

I asked him about population growth. Currently the population behind the rift is nearly three hundred people per square mile and will double in twenty years. So what happens at six hundred per square mile? People can continue to hack deeper into the wilderness along the Malagarasi, but most will end up in an urban concentration and contract AIDS. In the small villages along the rift, only one to two percent of the population is infected with AIDS so far. Maybe five percent of the truck drivers. It is an urban disease—in Dar it may be above thirty percent—and the excess population here will move to town eventually.

The problem, as Dr. Rosling observes, is that there are no more Minnesotas. In the 1860s, Sweden became increasingly urbanized, social mores broke down, and one-fourth of all Swedish hospitals were devoted exclusively to the treatment of syphilis. But then most of the excess population was able to emigrate to Minnesota and resume a rural lifestyle.

At this point, I told Dr. Rosling of my home base in Minneapolis, and we shook hands. Small consolation to him for not seeing any chimps today.

It is half past five and still hot; the birds and bugs are all singing, and the bats will soon join in. As we walk through the forest, the smells change startlingly from flowering trees to rotting flesh. Green leaves, red soil, everything's pretty much the same. The lake is still the world's greatest bathtub. This is more your home than mine—you ought to come out yourself next time.

I assume you will have heard my real news via Barbie by now.

Love,
Craig

MONDAY, 18 NOVEMBER

When Jane Goodall first came to Gombe in 1960, the chimps were so shy that she could only watch them from great distances. Each day she carried a telescope to the top of a narrow ridge top and caught fleeting glimpses of the chimps in the valleys below. She soon discovered that chimps eat meat and use tools, but she rarely could observe their behavior in any detail. Then an old male named David Greybeard came into her tent one day and stole a banana. She put out more bananas to encourage him to return, and he brought back a friend. Eventually she provided whole bunches of bananas for the entire chimp "community" that began to feed each day in the small clearing around her camp. She christened the group the Kasekela community, one of three chimp communities in the entire park.

Although Jane could now watch the chimps in their midst, the situation was highly artificial. Dozens of chimps crowded together for whole days at a time. In the absence of provisioning, a chimp community is usually scattered over a wide area feeding on fruits, nuts, and young leaves. Starting about twenty years ago, the Kasekela chimps received bananas only once a week, just enough to keep them coming through camp occasionally but not enough to interfere with their normal behavior. These days, most observations are made in the forest, but the chimp camp still serves as a landmark for both the Kasekela community and the researchers.

The feeding station is about half a mile up from the lake and a hundred yards above Kakombe Stream. It consists of two small metal buildings in the middle of an open square of cleared forest. One building, PP for Pan Palace, is connected to a long subterranean trench where the bananas were once delivered to specific individuals by a complex system of levers and doors. The delivery system has broken down, and now the bananas are dispensed by hand. The other building, LL for Lawick Lodge, is the observer's office. African observers maintain a constant vigil ("attendance record") for six hours before handing over the watch to a co-worker. The observer writes down details about each chimp that enters camp.

At certain times of year, the chimps feed on ripe fruits high up in distant valleys, and don't come to camp for days on end. In seasons when most of the community is feeding in nearby palm groves, the animals frequently use camp as a meeting point. Chimps can be extremely difficult to find in the forest, so the feeding station is a good place to wait for someone to show up.

Anne's chimp research has focused on the social and physical development of adolescents, and she needs their weights to measure their growth. Faster growth rates confer important reproductive advantages. Young females do not start cycling until they reach a critical body weight, and young males are not taken seriously by adults until they are large enough to dominate the females and the withered old males. But no weights have been collected in about five years.

To resurrect the system, we have suspended the scale from a chain that is hung like a clothesline between the roof of LL and a nearby tree. A rope dangles from the scale almost to the ground, and a bucket has been tied to the rope just below the scale itself. Our efforts yesterday were fairly frustrating, a bit like throwing our ox hearts out to the lions, but at least the chimps are madly keen on bananas. Perfecting the system is merely an engineering problem: we just have to arrange the scale so that the only way to the bucket is to climb the rope. The real problem is waiting for the chimps to come to us. Only one or two chimps may pass through camp in a day, but I'm grateful for the excuse to sit still and do nothing for a few days.

The sky is almost cloudless. Less than a mile away, the grassy peaks of the rift glow emerald in the morning sunlight. The upper tributaries of Kakombe Stream expand down along the valley walls and merge into the lush forest two thousand feet below. Around camp, forest birds sing invisibly in every direction. Then a series of pant hoots echoes through the valley: waaaah waaaah waaaah WAAAAh WAAAAh waaaah waah. A chimp has arrived on the ridge top to the south of camp.

Each time the chimps cross from one valley to the next, they enter a new sound domain. Like lions, the chimps are often separated from their companions by long distances, and they call loudly to locate each other. Also like lions, male chimps defend group territories. They need to recruit their distant companions and intimidate their enemies.

After half an hour, a lone male chimp walks silently into camp. Eslom is the observer of the day and he tells me that it is Wilkie, a chimp that was only a baby when I first came out here. Now Wilkie is twenty and in his prime—the top-ranking male in the Kasekela community. Eslom places the banana in the bucket as Wilkie patiently waits. Eslom steps back to watch the scale, and Wilkie climbs up, removes the banana from the bucket, and feels around inside for more. The scale settles on ninety pounds. Ninety pounds of solid muscle built for climbing and clinging. Pound for pound, a chimp is three times stronger than a man.

Wilkie carries his prize down to the ground in his mouth, then eats slowly in the shade of a bush. He waits for a few minutes, listening, then departs as quietly as he came, continuing on to the north. I ask Eslom where the rest of the males are today. He says they nested in Kahama valley last night, near the southern end of their range, and that the B-record is with them again today.

In addition to monitoring the chimps that pass through camp each day, the staff also follow them wherever they may go. On a duty called B-record, a pair of "followers" records the behavior of a selected animal for the entire day. There are ten chimp followers altogether. Each man conducts one or two all-day follows per week, so at least one chimp of the Kasekela community is under observation at all times. The chimp followers have the highest prestige of anyone here, for everyone knows that Gombe is famous only for its chimps. The chimps may travel through five or six valleys in a single day, and you have to be tough to keep up. The tougher the follow, the greater the bragging rights.

Unfortunately, Hilali, the headman of the chimp study, is not at Gombe this week, and I am sorry to miss him. Hilali is a Waha, born in one of the nearby villages. The Waha know these valleys and plants better than anyone. Most are Muslim, thanks to the Arab slave trade, and the rest are Christian, thanks to Dr. Livingstone. They will accept Western medicine, but they also believe in magic. Every year someone seems to fall into a fatal funk after being hexed by a vengeful neighbor. The Waha live on fish from Lake Tanganyika, but most don't know how to swim.

Like others from his tribe, Hilali is a small man, but he has the most remarkable, imperious dignity. As a lowly baboon student in the old days, I never had much to do with Hilali, except after returning from my first trip to the Serengeti in late 1974. Anne and I had gone to visit old Gombe friends, David Bygott and Jeannette Hanby, who had just taken over the long-term lion study at Seronera. We both marveled at the wide open spaces but had no idea then that we would eventually come back to study the lions ourselves some day.

The four of us headed back to Gombe, piling into the Bygotts' Land Rover and driving west from Seronera toward Lake Victoria. The roads were flooded and we decided to drive north cross-country until we reached drier land. Whenever we reached a river, I would hop out to look for a crossing. While running down one sandy gorge, I stepped on an acacia branch and drove a long thorn through my sandal, deep into my heel. Anne, David, and Jeannette all took turns with tweezers and pocket-knives trying to pull it out, but to no avail. The thorn was so rotten it broke off under my skin. Two days later we arrived in Kigoma, and I was quite concerned about going back to wet, bacterial Gombe with the thorn still embedded in my foot.

Emilie, the Dutch administrator, covered my wound with some black sticky goo that she always used to treat horses for hoof infection (Why did she have this stuff at Gombe?). I was in pretty serious pain after another day, and Emilie said, "Why don't you get Hilali to pull the thorn out? He's the thorn *fundi* (specialist) of the staff camp."

We met Hilali outside the Mess, and he instructed me to lie on the ground on my stomach. He looked at the wound, then asked for a knife.

"What kind of knife?"

"One that is very sharp, about six inches long."

"Why do you need such a big knife?"

Hilali didn't answer. Emilie returned from the Mess with a carving knife.

"Good," said Hilali.

He cut off two stout twigs from a nearby bush, whittled off the bark and sharpened one end of each stick. Finally satisfied, he came back, took my heel in hand, and drove one of the sticks into my foot right next to the thorn. I have a fairly low threshold of pain, and I suppose I probably winced at this point. In fact, I probably jumped a bit.

Hilali instructed Anne and Emilie to sit on my back to keep me from moving, then returned to work. He reinserted the first stick, then jabbed in the second. I was quite glad to have two people holding me down at this point. They somehow managed to keep my foot still long enough for

Hilali to use his probes like chopsticks and yank the thorn right out of my heel. He handed it to me and left without a word.

These days Hilali is a proper *mzee* with a dozen children and, according to Anthony, he is loftier than ever. He still hauls himself up these steep slopes in the heaviest rains, even though he must be in his fifties by now. He recently told Anthony: "We are born, we work, and we die. I have my children and now I can die."

Eslom must be pushing fifty himself by now. He too has been at Gombe for many years, but he doesn't have the same self-possession as Hilali, or the same stature in the community. He has only one wife and three children. Eslom is curious about all the students he used to know—where they are now, what they do, whether their parents are still alive, and, more to the point, whether they're rich.

I have a pretty good idea what he is driving at, but decide to play along, compressing my reply into a few summary sentences. Our own demographic landmarks sound much the same as those of any other species: movements from one community to another, reshufflings of relationships, births. Although some of us have lost parents, none of us have died so far, but the study is still too short to say how long we'll last.

But Eslom is not interested in these abstractions, he is concerned with the here and now. He wants to retire, he is getting old, his salary is low, he needs money. "Tell me, Hi-guy, how can I contact all the old students. How can I raise money for my retirement?"

Promising to talk to Anthony about raising his salary, I make a hasty farewell and run off to join Apolly and Christine who, fortunately, have just walked through chimp camp. They are going up to the Dell to look for C troop. The Dell is a boggy area and a good spot to catch bilharzia, or schistosomiasis, a disease caused by a worm that alternately infects snails and primates.

C troop always has a few old females that look exceptionally grotty. This morning, Christine had given a chocolate laxative to a hairless female named Penelope. Penelope is not even twenty yet, but looks much older. She climbs onto a low branch, chews on a palm nut, shifts her scrawny bottom half an inch, and lifts her rat tail. Plop. Gloved hand quickly scoops up the droppings. "Oh, look, a worm!"

I am quite shocked at the sight of Penelope. When I first worked with C troop I thought the hairless old females were just very old. But Penelope is younger than many of the females we saw with D troop the other day, and they all had sleek coats and seemed to be in the best of health. Perhaps Penelope has too many worms, or perhaps C troop inhabits an impoverished patch of forest.

Apolly has assigned his entire team of baboon followers to help Christine. She has given each man a plastic bag filled with specimen tubes, rubber gloves, and tongue depressors for scooping up the dung. They have already collected over seventy samples, and the chocolate laxatives have gone down as a treat. Now, no matter how badly things turn out with the lions, Christine will have all the specimens she can handle from the baboons. In this business, one out of two isn't bad.

I'm very happy Christine can manage without me. Following baboons around through the forest for two years, slipping down wet slopes, crawling through thickets on my hands and knees, I invariably came into contact with more shit than I care to remember. However, I do owe a lot to baboon shit. In May of 1975, Anne and I took a long holiday to recover from a bad case of hookworm. On the nineteenth we were in a hotel on the Kenyan coast, taking our worm medicine. If we had been at Gombe that night, we would have been kidnapped with the others.

Looking again at Penelope, I am reminded of another pathetic old female from C troop named Sheba. Sheba was a real bag of bones who eventually died down by the staff camp. But her son, Samson, was just reaching his prime when I first arrived in 1972, and he changed the course of my life.

I originally came to Gombe as an undergraduate assistant and only expected to stay for six months before entering medical school. I had been reluctant to leave all my college friends in California and was somewhat intimidated by the graduate students and postdocs in the Gombe research community. Climbing through thickets with Apolly each day was difficult and boring. It was hot, there were too many bugs, the place was crawling with snakes. It seemed impossible to recognize individual baboons—the nondescript faces of juvenile baboons would flicker through my mind each night as I tried to fall asleep. Why would anyone waste their time on the dull routine of watching monkeys?

But then, one morning by the Mess, I found Samson sitting in a tree near B troop. He was alert and nervous, intent on the strangers about a hundred yards away. One of the B troop males suddenly raced down the beach and chased Samson back into the forest. I tried to follow, but the two baboons streaked up the hill like lightning. That evening, I found Samson back home with the rest of C troop, but next morning he was hovering near the Mess again. He lurked behind trees, more cautious than the day before, but he seemed determined to stay close to B troop.

Apolly had told me that all male baboons moved from one troop to another sometime during their lives, while the females remained in the troop in which they were born. But Samson's behavior just clicked with me at that stage in my own life. Here was a male at an equivalent age to myself about to make the most amazing journey in his life—he might only move a quarter mile from home, but he was about to leave all his childhood friends and family and set off into the unknown.

But why? Why leave home? Why not just stay with mom and the family? Why did each and every male baboon go through this self-inflicted rite of passage? No one tells these animals what to do. Something must intensely motivate them to move on. Samson was facing violent opposition from the other B troop males. He could go home to C troop and no one would ever bother him. But he was determined to get into B troop at all costs. Why?

I was hooked. I asked Jane for permission to stay for an extra three months, gave up my place in medical school, and became a research student in England the following year. Then I came back to Gombe for another year to answer just that one question. Why?

Eventually I watched more than a dozen males move from one troop to another, each making that night journey, that voyage from childhood to maturity. Some had an easy time; others spent months surveying a variety of neighbors, risking life and limb. A male baboon has longer canine teeth for his size than a lion, and several of my males were badly bitten during their move. In the end, Samson had to give up on B troop and join another troop elsewhere. But he couldn't be stopped. Samson, like all the other males, saw something in the new troops that obsessed him. They might change destinations, but the trip had to be completed.

The attraction turned out to be sex. Although a baboon troop may contain forty or fifty animals, it is just one large family, and males and females that grow up in the same troop simply don't excite each other. Females prefer males that enter their troop as adults, and males are willing to do battle only for females that they meet in new troops. Like many other animal species, baboons show an aversion to incest—they experience sexual indifference toward the individuals with whom they have grown up—and the lust for sexual novelty inspires young males to abandon their home troop for another. Samson wasn't looking to find casual friends in a new troop, he was seeking new females—all those fresh faces, all those exquisite red bottoms.

A few days after seeing Samson, I danced with Anne for the first time. The tables in the Mess were shoved against the wall, the night was warm, and the stars were shining outside. The Jefferson Airplane sang:

> If you feel like china breaking
> If you feel like laughing
> Break china laughing
> Break china laughing laughing laughing

I whispered in her ear, "Sweet nothing." She whispered back, "Sour something." Anne came to Gombe shortly after graduating from Oxford and went to America as a graduate student during the same years that I was enrolled in England.

By a rather curious coincidence, Anne had become intrigued by intergroup migration in chimps. But in chimps it is the female that moves from one community to another. Young females experiencing their first estrus start searching for males in nearby communities. They may return home between cycles at first, but most females ultimately settle in an adjacent community. In contrast, male chimps remain in the community of their birth. As a result, chimpanzees have a very unusual kinship structure. In lions, baboons, and most other group-living mammals, female companions are close relatives. But in chimps, only the males of the same community are typically close kin.

This is one of many characteristics that reveal the chimps' close evolutionary relationship to humans. Traditional human societies usually exchange females between groups. The males possess the wealth and stay close to where their fathers were born.

Now I'm back here again all these years later, and we are asking another question. Why should baboons and chimpanzees, two different species that live in the same forest and eat more or less the same food, show such striking differences in their dispersal and kinship patterns? What makes males get up and go in one species but stay put in the other?

This question may seem fairly abstract. But male kinship is associated with another phenomenon that we ignore only at our peril. Kinship is the glue that binds males together when they go to war. We few, we happy few, we band of brothers.

TUESDAY, 19 NOVEMBER

Gremlin carries Galahad up the rope. Together they weigh ninety-seven pounds. Gremlin sits on the ground, and Galahad begs for a bite of her banana. Gremlin keeps her mouth shut, refusing to share. Galahad wanders off a few yards from his mother, climbs a low tree, and dangles from a branch. Seeing our chance, we put a second banana in the bucket. Gremlin quickly scoots back to the scale and climbs the rope by herself. Eighty-one pounds; Galahad must weigh sixteen.

Female chimps are relatively unsociable animals, traveling only occasionally with other adults and preferring to stay alone with their dependent offspring. Galahad is not yet three and won't be weaned until he is five. He'll stay with mom almost constantly until he is nine or so, then hormones will take over. At puberty, young males start their apprenticeship with the adult males. Adolescent females venture off during their first sexual cycles. Spring torrents.

Gremlin carries Galahad a few hundred yards out of camp and climbs the fronds of an oil-nut palm. Palm nuts are covered with orange and purple flesh that is very high in fat and protein, and tastes something like an avocado. The palms grow next to permanent water and produce fruit year-round. They are especially valuable to the chimps during the dry season when other foods are scarce.

Females rarely forage with other females unless a particularly large fig tree is in full fruit. Each adult female has her own "core area." Some females center their activities here in Kakombe valley, others live in the next valley north, still others to the south. Some females stay close to the lake, while others remain closer to the rift. We know little about the relative quality of each female's range, but it seems likely that a high-ranking female such as Gremlin has a core area that is relatively rich in year-round food sources.

Although most young female chimps move into new communities at Gombe, several Kasekela females have remained in their natal community and have continued to associate with their mothers as adults. Anne wants to know if the mothers of these stay-at-homes have especially rich core areas. A young princess might be reluctant to give up her mother's palace to marry the nobleman in the next village—she might not find such a rich estate in the new community. In chimp terms, the most likely source of local wealth is palm trees. Palm nuts are the staple diet of female chimps throughout the year and are most abundant in the lower valleys. Tomorrow I have to start clambering around the ridge tops, mapping the location of each palm grove in the Kasekela range.

Gremlin is concentrating on her lunch, dropping discarded palm nuts to the ground. Galahad hangs from the palm fronds and otherwise tries to keep himself amused. He is already interested in the adult males, but it will be years before he leaves the security of his mother to tag along with them.

Male chimps are much more gregarious than females. Once a week or so, the dozen males of the Kasekela community join forces and police the edge of their range, looking for neighboring chimps. While they are together, they may take the opportunity to catch a colobus monkey or infant baboon. The males maintain a complex set of dominance relationships. The top male usually has greater access to estrous females and maintains his rank with cunning aggression and a shifting set of alliances. But male chimps also form long grooming chains, with each male intently gleaning the hair of the male in front. They also have a curious way of greeting each other after a separation, by screaming loudly and fondling each other's testicles. It is the dramatic behavior of the male chimpanzees that captures most people's attention, the males that provide the guts of the day-to-day soap opera that makes for such compulsive viewing.

Jane studied the Gombe chimps by herself for several years but started to bring in a steady trickle of European and American assistants during the '60s. The assistants recorded every nuance of the animals' behavior, and the work was brutal. During the height of the banana feeding, the chimp community spent the entire day in camp, and there was constant action—grooming, fighting, mating, infants nursing and playing. Battles raged between chimp and baboon over the bananas. The assistants observed the chimps constantly from six o'clock each morning until seven o'clock at night, then stayed up until two or three o'clock in the morning typing out their tape-recorded notes. Few people lasted more than a couple of months.

As Jane's fame grew, more scientists arrived. A student from Berkeley started the baboon study; a Cambridge student studied the red colobus monkeys. More chimp students and assistants arrived from Britain and the United States. Then came a rather short-lived link with the Friends World Institute. The FWI people were wealthy American undergraduates who came straight from the adolescent subculture of the late '60s and gave Gombe the reputation of the free-love capital of East Africa. Besides taking notes, the FWI students were supposed to photograph striking examples of chimp behavior and send the films to be developed in Nairobi. They did, but they also sent graphic images of their own sexual behavior.

In 1968, the intensity of banana provisioning was drastically reduced, and observers started to follow the chimps out of camp for the first time. Although no one yet suspected it, the Kasekela community was already starting to divide. About half of the males spent more and more time in the southern valleys, and one student in particular made an intensive effort to monitor them, going out alone to every far-flung valley to which the chimps might travel. Then, one day, she didn't come back.

Ruth was studying the males. She was intrigued by their individual personalities, their Machiavellian manipulations. As the southern males came to camp less frequently, she had to take advantage of their every appearance to follow and observe them. By following the males to their sleeping trees, Ruth could resume her observations the next morning. Otherwise she might not see them again for a week. After Ruth failed to return one night, search parties set out to scour the park. It took six days to find her.

Ruth had fallen down a high cliff in Kahama valley. It was impossible to know exactly what had happened. Chimps sometimes charge straight up steep rock surfaces, and Ruth may have tried to follow directly behind, rather than take a safer route off to one side. There was a beehive nearby; perhaps she had been swarmed while climbing the cliff.

Ruth's death led to a new policy. Henceforth, everyone following the chimps must work in pairs. Each student would be allowed to work outside the feeding station in the company of a Tanzanian assistant. Hilali and Eslom were hired shortly thereafter, and Apolly came in 1970. Earlier, Jane had employed the local people only as domestic help, and initially the Tanzanian field assistants were little more than trackers or bodyguards. As the years passed, the Africans were given more and more responsibilities. By the time of the kidnapping, they were able to carry out the fieldwork as well as anyone. Foreign students were no longer permitted at Gombe after May 1975, but they were no longer essential to the project either.

But in 1970 Jane became a visiting professor at Stanford, and the students at Gombe started to discover aspects of chimp behavior that could never have been observed at the feeding station. Up to this point, chimps had seemed to lack the vicious streak so common in other animal species. There were fights within the community, to be sure, but they seemed like minor domestic squabbles heightened by the artificial circumstances of banana feeding.

Then, in 1971, David Bygott followed a group of Kasekela males to an adjacent valley where they encountered an unknown female with a small infant. The males attacked the mother, grabbed her infant, and ate it. Our peaceable relatives turned out to be cannibals.

The chimps' "peaceful" reputation had rested entirely on observations of their behavior toward members of their own community. The other two communities in the park were extremely shy and could not be tamed with bananas. But by the early '70s the Kasekela community was clearly dividing into two groups. Six males and three females centered their range in the southern valleys. The remaining eight males and twelve females ranged closer to the feeding station. Intercommunity relations could be studied for the first time.

Encounters between the two male groups became increasingly hostile over the next three years, though coexistence still seemed possible after the original territory had been partitioned. Then the northern Kasekela males launched a series of invasions into the southern valleys. Whenever they located a southern male by himself, they would quietly stalk him, surround him, and launch a wild, vicious attack. Several of the males would work together to pin their opponent down while their companions charged in to bite and pound with absolute ferocity. These attacks left each victim mortally wounded, bleeding from huge gashes and with broken limbs. Emilie came back to the Mess in a state of shock one day, pale and horror-struck. She had just seen the northern community attack the old male Goliath. They had literally torn him apart.

By the end of 1974, all but one of the southern males had been murdered, and the purpose of the attacks became clear. The northern males had completely annexed the southern community's range and claimed two of their females. Agincourt was theirs.

But by the '80s, the victorious warriors had dwindled to just five males and were themselves threatened by a neighboring community. The Kalande community had always been too shy to census and remained at least three or four valleys south of camp, but now the Kalande males moved north and met little resistance from the enfeebled Kasekelas. Thus emboldened, the Kalandes invaded the feeding station and assaulted a Kasekela female.

In the following two years, several young Kasekela males reached maturity and the Kasekelas started to regain their former glory. The ten Kasekela males now dominate the park, traveling with impunity wherever they wish, and the Kalandes have retreated to the south. To the north, the small Mitumba community is cramped along the park boundary.

Warfare may be a manly art among the chimps, but among the baboons the females are the custodians of the land. Female baboons remain in their mothers' troops and inherit their range. Females form rigid hierarchies and are often unpleasant to each other, but they will work together to defend their range against invasion by neighboring females.

As opposed to male chimps, male baboons are mere soldiers of fortune that move from troop to troop, looking for wives and mistresses. A male's companions are his rivals rather than his partners. They compete against each other for mating opportunities, and fighting is often intense. Although they sometimes act in concert to prevent additional males from entering their troop, male baboons show no organized hostility toward males in other troops.

Why do chimps form groups of marauding males instead of acting individually like male baboons? Richard Wrangham studied this question during the early '70s and felt the answer lies in the solitary nature of female chimps. Females are scattered widely, and a single male could only defend a territory large enough to encompass the ranges of a few females. By banding together, males are able to defend a much larger area, and thus a pair of cooperating males might gain access to more than twice as many females than either male could defend by himself.

Male baboons, on the other hand, must contend with a different distribution of females. Female baboons forage together in groups too large to be monopolized by a few males. Rather than trying to sequester the females, male baboons merely flit from troop to troop, skimming off the cream.

The happy band of male chimps that had annihilated the southern community was composed of brothers and cousins. Why is kinship so important in promoting group cohesion in male chimps? Why do they spend their lives in the community of their birth?

Our lion studies may shed some light on these questions. Like male lions, unrelated male chimps should cooperate only if each male would thereby father more offspring. And like coalitions of male lions, bands of male chimps do not divide matings equally. In many cases the top-ranking male chimp mates more often than his companions. Ranking third or fourth on the totem pole would bring few rewards, except for the genetic advantages of assisting brothers to produce more nephews. In theory, coalitions of close kin can grow larger than those composed of nonrelatives because, as in lions, bands of brothers will continue to cooperate in groups too large to provide a fair share of matings to each member. Brotherhoods are able to overpower any smaller group of nonrelatives and thus gain access to greater numbers of females. Eventually, all the other males in the population have to cooperate with their brothers, too.

With their large litters and synchronous breeding, female lions can easily produce large cohorts of young male relatives. These cohorts provide most young males with a preassembled coalition when they are ready to set off and conquer the world. Female chimps, however, produce only one offspring at a time. In order to cooperate with their male kin, young male chimps have to stay among the males with whom they have grown up. They have to stay within their patriline.

This framework makes the story sound very tidy, but there are many loose ends. While it is clear that male chimps are most cooperative when battling neighbors, the precise evolutionary advantages of male cooperation have not yet been measured. No one knows if males in bands of six actually father more offspring apiece than, say, males in bands of four. Only one chimp community has been studied in detail at Gombe. The other communities are mere shadows at either limit of the park. We can measure the precise reproductive advantages of cooperation among lions because we know the histories of so many different prides and coalitions. Chimp research has inevitably, and rightly, taken a different focus. Concentrating on the detailed behavior of each individual, Jane Goodall has stressed the personalities of Flo, Fifi, Passion, Goliath, and the rest. But no individual animal lives in a vacuum, nor does any single group. Individual behavior can only be understood when viewed from the perspective of the population as a whole.

What you do in your own group has important consequences for your struggles against every other group. We have returned again to the same old movie: us against them; we all pull together on the battlefield.

We all do.

WEDNESDAY, 20 NOVEMBER

The surface of the lake is calm and warm. I stop swimming for a moment, tread water, and feel the cold down below. A bee buzzes around my head, looking for a place to land. Splash, splash. "Buzz off!" But I'm the only landing pad within reach. I dive underwater and head further out into the lake. The water is clear, but too deep to see the bottom.

On the beach in front of me is the "Cage," the metal house where Jane lived with her son Grub when he was young and needed a playpen safe from the chimps. Nearby, the Mess is hidden behind a wall of vegetation. Along the beach to the north, Jane's new stone house is barely visible behind the shrubs and vines. I swim farther from shore. The mountains of Zaire stretch across the western skyline, flat and black beneath the hazy afternoon sun. To the south, only the tops of distant thunderstorms are visible above the horizon.

The bee is back, but I can also make out another dull droning sound, a boat engine. A water taxi is churning along from Kigoma, heading north toward the Burundi border. It is a wide, wooden-hulled boat so crammed with people that the gunwales barely clear the waterline.

Water taxis take three hours to reach Gombe from Kigoma. The boats leak, and the passengers have to bale constantly. People make their own small contributions to the churning swill: children pee in it, adults spit in it. Some water taxis are shaded by corrugated iron roofs that trap an atmosphere of dried fish and tropical body odor. There is almost always someone on board who is coughing or sneezing, and around this lake a cough can carry tuberculosis or meningitis.

I once sat in a water taxi between Anne and a small, timid child. The child seemed unusually quiet, and I assumed that he was merely nervous to be sitting so close to an *mzungu*. Halfway from town, he turned and vomited onto the bench. He had thrown up a large wad of white worms.

The water taxi passes by about a hundred yards away. Occasional voices rise above the garbled roar of the engine, which suddenly hiccups a few times. The sputtering is harmless, but it sends me swimming back to shore.

On 12 May 1975 a water taxi churned past the research center. The engine exploded when it reached the next valley to the north, and the boat capsized about two hundred yards from shore. Most of the passengers could not swim. Jane, Anthony, Helen, and nearly all the other Gombe students were down by the lake at the time. They didn't see the accident, but they heard the commotion and ran along the beach. It was too late. Dozens of people had drowned, and the students could only help bring bodies to shore.

Anthony is at the park offices this evening, going over the accounts and dispensing medicine; Christine is still out collecting samples from

B troop; Craig Stanford is off somewhere with his colobus monkeys. It is my turn to cook supper. We are no longer in a sparkling white kitchen built by Texas A&M in a clean, arid environment. There is no refrigerator, we have no rosemary; we have very little of anything. We are on the shore of one of the largest lakes in the world, but we have no fish.

Gombe National Park does not extend down to the beach. For two weeks each month, fishermen are permitted to live within one hundred yards of the lake in palm-frond huts. They paddle out each night in their canoes, casting their nets for the silvery *dagaa*, which they dry on the beach, and for larger fish, which they sell fresh or smoked. Fresh *dagaa* taste like sprats, and *tilapia* are like chicken. But the moon is too bright tonight, and all the fishermen have gone home.

Jane's kitchen is dark and dingy even in the middle of the day. As evening falls, I light a couple of old hurricane lamps, really primitive kerosene lamps like the one kicked over by Mrs. O'Leary's cow. The fruit and vegetables are stored in a cabinet with screen doors that keep the bugs out, but ants have gotten into the bread, the cashews, and the open tin of jam. Most of the vegetables are moldy. Well, let's just chop them up and make the best of what's here.

Digs arrives to make *chapatis*. Digs spends most of his time with the Mitumba community, the small group of chimps along the northern park boundary. No one had succeeded in approaching the chimps outside the Kasekela community until the followers located two Kasekela females that had transferred to Mitumba. The two Kasekela females are so accustomed to human observers that the Mitumba males are starting to lose their own fear.

Digs decides to experiment with wheat flour and corn meal—Mexican *chapatis*. We spend ages rolling the dough and stoking the open fire on the kitchen floor, trying to keep the pans sizzling. The hurricane lamps cast a vague orange glow, illuminating only nearby objects. All else is lost in the dark.

The others return from their day's labors. Craig Stanford spent the day in Linda valley, where the chimps tried to catch a colobus but failed. Christine has collected nearly two hundred fecal samples. Anthony has been teaching the new administrator to deal with the monthly wages. The new man is the son of Rashidi, the courtly *mzee* who was the first headman of the Gombe staff village.

Beer has appeared, beer from Burundi in large brown bottles with the label spray-painted directly onto the glass. Anthony passes around glasses of lime juice and Konyagi, Tanzania's answer to vodka. It is clear and rather flavorless, but it gets the job done.

Anthony selects a tape of country and western singers performing songs by the Grateful Dead. Conversation grows louder, the *chapatis* start to look good, yes, another Konyagi-and-lime sounds great, people are bouncing in and out of the kitchen and setting the table. A cheerful twanging voice pours out:

> Sugar Magnolia, blossom's blooming
> Head's all empty and I don't care

"More Dead, Dr. Collins, I must have more Dead."

Anthony replaces the short-lived Tanzanian batteries, and hands me a new drink.

"While I was in Kigoma market yesterday, looking for eggs, I turned around and thought I saw Nasibu."

"Nasibu!"

"Yes, that's how I felt."

Anthony turns the tape over and the music resumes with "Dark Star." The room suddenly goes dim, and my body is seized by fear. I am swallowed by the darkness, swept into the past.

I need music in these far-flung places; I need to have something running through my head. My first trip to Gombe in 1972 was like an experiment in sensory deprivation after a lifetime of American overindulgence and incessant stimulation. I wasn't prepared for the shock of these silent foreign shores. My friends and I frequently went to the Fillmore Auditorium in San Francisco. The Grateful Dead loomed large in local folklore. To us, "Dark Star" was to rock music what the Kreutzer Sonata is to chamber music or *Götterdämmerung* is to opera. Quietly intense and well over twenty minutes long, a world in itself.

Everyone else at Gombe lived in huts up in the forest, but I lived in the Cage to be closer to the baboons. My quarters were rather damp, and I had to share space with the research center's boat engine. But I had a clear view of all the fishing boats at night and the Burundian refugees that filed by during the day.

Friends back at Stanford sent occasional tokens of civilization. One sent a McDonald's cheeseburger and labeled the contents "One toy cheeseburger." It arrived three weeks later, a rather beautiful malachite green. Another friend sent out an unmarked cassette tape. I immediately went to the Mess and discovered that it was "Dark Star."

But the recording was awful. The sound wobbled and faded in and out. I spent hours after supper trying to rebuild the cassette so that it would somehow sound right. I finally admitted defeat and went back home to the Cage just before midnight.

I went to unlock my door, but it was already open. Thieves had broken in and stolen our boat engine. Everyone told me I had been lucky. If I had gone back to the Cage during the theft, I would have been attacked with *pangas*. A defective tape had saved my life.

The next day, Jane decided to hire a night watchman. It is common knowledge that local people will steal from anyone who doesn't have a watchman. A watchman need not be big and fierce. A token watchman would do. Like Kisiri.

Within days, Jane hired Venas. I never knew whether he was called Venas from birth or whether it was an ironic nickname given to him in adulthood. He simpered and always spoke in a high falsetto—it was especially peculiar to hear him greet me with a piercing, "Hi, guy." He was so afraid of walking around at night that his voice trembled in the dark. I couldn't imagine how he had ended up as a watchman; maybe he could scream like a siren.

At the time, Apolly's chief assistant in the baboon study was Nasibu, a charming Zairian of dubious character. There was talk that he had been behind the theft of the boat engine. There was a lot of illicit traffic back and forth across the lake, and he could have disposed of the engine in Zaire. Once he became the focus of suspicion, Nasibu started drinking heavily, and alcohol made him surly. He was powerfully built, and people began to fear him. He was fired a few months later.

In 1973 I went back to Stanford for a few months, then enrolled as a graduate student at the University of Sussex. After we returned to Gombe in 1974, Anne and I moved into one of the metal houses up in the forest.

Chimps and baboons foraged in the palm trees in front of our verandah. Birds babbled from dawn to dusk. Bush pigs snuffled around outside each evening and crunched loudly on palm-nut kernels that had been rejected by the chimps and baboons. Every night, the fruit bats would squeak, the frogs would chirp and croak, and Venas would make his rounds with his flashlight and *panga* and shout "Hi, guy!"

In April 1975 Anne and I began to plan for a holiday. We had originally intended to take a steamship from Kigoma to Bujumbura, the capital of Burundi, and then journey onward to the Heart of Darkness. We had both lived at Gombe for years and had never been to Zaire. How could we even claim to have been in Africa if we hadn't been to Zaire? Theresa had had a wonderful time in Zaire. The people, she said, were all so friendly.

But then we began to hear reports of political unrest in Zaire, in the province just across the lake. The boat service from Bujumbura was terminated. There was no way to get across. Things might calm down in a few months, but in the meantime Anne and I decided to go to the coast of Kenya instead.

On the nineteenth of May, in the middle of the night, a boat carrying forty armed men came roaring across the lake from Zaire. They landed near the staff camp at Gombe and demanded to be taken to the *wazungu*.

The invaders grabbed the headman, Rashidi, but he refused to talk. They hit him over the side of the head with a rifle butt, bursting his eardrum. Still no answer. They saw Venas's flashlight as he came back from his rounds. Where are the *wazungu?* Venas wouldn't talk. They began to beat him. Not a word. Venas, the token night watchman, the laughingstock—Venas would not betray the students if it cost him his life.

Marching down the beach, the invaders found the Mess and the Cage, but no one was there. Jane was in her new house, but they couldn't see that far along the beach.

They were just about to give up and go back to their boat when they heard the sound of typing from above the cliff. Emilie was up late finishing off the accounts. She had come to Gombe as an administrator, but she watched the chimps whenever possible and often worked eighteen hours a day.

The Zairians followed the path up from the Mess to Emilie's house, surrounded her hut, then charged in and grabbed her. She put up a terrific fight, but she was hopelessly outnumbered.

Emilie's hut was only a hundred yards from the junior warden's hut. Ettha Lohay heard Emilie's screams and ran out to help. She saw several men drag Emilie out of her house; more men were heading toward her along the trail. Ettha threw off her white nightgown and dashed twenty yards into the forest. She reckoned that no one would see her black skin against the forest night.

As soon as the invaders had passed, Ettha ran to another trail and circled toward the rest of the houses in the forest, telling the students what had happened to Emilie and sending them deeper inland.

After they had passed Ettha, the invaders entered first her house, then Anthony's. He was on holiday in southern Tanzania.

Then they found Steve's house. Steve put up a fierce struggle. His angry shouts brought two more Stanford students, Barb and Carrie, to see what was the matter. They both walked straight into the hands of the kidnappers.

The invaders subdued all three students and continued along the path. They arrived at our empty house, looked inside, and shouted: "Where are these people? Why aren't they here?" They went to one last house, but the student had already been warned by Ettha and had fled into the night.

The kidnappers collected Emilie and the three students and broke into the storeroom by the Mess; they wanted to take plenty of supplies for the *wazungu*. Where they were going there would be no Western comforts.

The invaders seemed to have some idea where everyone lived; they seemed to be acting on inside information. Several of the staff said they recognized one of the invaders. They said it was Nasibu.

THURSDAY, 21 NOVEMBER

I walk up from the Mess on the same path followed by the kidnappers. The forest is still dripping from an early morning rain. *Siafu* (safari ants) have created a system of highways that crisscross the trail. Millions of tiny red feet have churned the soil into a series of half-inch embankments that disappear into the forest on either side. At my approach, grey triangles lying scattered on the ground suddenly rise up and unfold into sky-blue butterflies before vanishing into the understory.

Here are the bare foundations where Emilie's house stood, there was Anthony's house, over there, Ettha's. That bare patch used to be their outhouse. It reminds me to watch my step.

A visitor once went to use that outhouse during an afternoon get-together. She came back laughing uncontrollably, then started to cry hysterically. While she had been sitting on the toilet, a large cobra had reared up between her legs and spread its hood. She'd hopped up onto the toilet seat, leapt over the cobra, then bounded out the door.

Here are the remains of Steve's house, then ours. The crumbling concrete slab is covered with creepers and fallen palm fronds. A few yards away stands a termite mound, massive, solid, and symmetrical, like a lost Mayan temple swallowed up by the forest.

The night following the kidnapping, the remaining students huddled together in the Mess. A brilliant white light shone briefly on the mountain slopes of Zaire. What could it mean? There were never any lights over there; the distant mountains had always been a dim, lifeless wall. At about midnight, a boat roared across the lake straight toward the research center. The terrified students fled into the forest beyond the chimp camp.

But the boat was just a water taxi, off course and lost in the night.

Several days later, the remaining *wazungu* were evacuated by water taxi, and after a day in Kigoma, took the train for Dar es Salaam.

Anne and I had reached the station in time to say good-bye. We spent the night at a guesthouse on a small hill above Kigoma. No one turned out their lights. No one knew anything about the invaders except that they had spoken French and had worn military uniforms. The fact that they had stolen supplies meant they must be planning to hold the hostages for a long time.

The next morning we flew to Gombe in a military helicopter. Apolly, Hilali, and the rest of the staff were all in a state of shock. Rashidi had a bandage around his head; Venas's face was swollen and bruised. Ettha was wearing her park uniform, still in charge but deeply anguished. Everyone was shattered, withdrawn. We all had lost friends to the kidnappers. Foreign students were no longer allowed to work at Gombe. This was our last farewell.

The police had given us a half hour to retrieve our data and the rest of our meager possessions. I packed my accounts of the baboons during their own fearful journeys. Anne searched drawers and shelves for her data; some of it was still up in chimp camp. What to do with all the mementos of our years in Gombe? Wooden carvings, polished stones from the beach, the orange-and-blue cloth that hung from our rafters—"*Oyee* Tanzania *Oyee*! (Hurrah, Tanzania, hurrah for ten years of independence!)" The abstract patterns had always danced before my eyes whenever I lay in bed with fever.

"*Hodi.*"

It was Steven the carpenter. Steven came from Arusha, and he belonged to the same tribe as Ephata. He had built Jane's new house and the Mess.

We exchanged distracted greetings and quickly resumed packing. Something seemed to be on Steven's mind. He wouldn't leave us, but he wasn't saying much either. We only had a few minutes; the helicopter pilot was an impatient man. Steven was in the way, standing next to our cupboard, staring underneath our bed. Finally, in exasperation I asked, "What do you want, *bwana*?"

He got down on his hands and knees, reached under our bed, and pulled out the near-empty bottles of liqueur that we had accumulated over the years.

"Will you be taking these with you?"

I laughed. "No, you can have them—but share with the rest of the village."

"Yes, of course" he promised theatrically, and then proceeded to open each bottle. He poured the Drambuie in with the whisky, then added the Cherry Heering and the Benedictine.

Expressing the profound depths of his sorrow one last time, Steven bade us farewell. He stepped outside, took a huge swig of his concoction, and staggered back down the hill.

Continuing on up the path though Plum-tree Thicket, I duck the low, looping vines that dangle from the overhanging trees, step over yellow fruits that have been dented by the teeth of a chimp and discarded as unripe, then push aside air-roots hanging from the canopy like thirty-foot strands of spaghetti. Black-and-white hairy caterpillars crawl on the leaves of a low bush; their tiny legs ripple along under a brush of poisonous spines.

Baboons rarely stay on the trails, so I used to have to crawl through these thickets and pull myself up the slopes between low branches while holding on for dear life, balancing precariously on stones or rotten logs, peering through narrow windows among the tangled branches, trying to see what Samson or Jonah or Hamlet was doing. Hidden in the thicket, the enraptured young male would introduce himself to a female swollen with unquenchable desire, terrified that an older male might suddenly appear out of nowhere and send him home, bleeding, to his mother.

Emerging from the thicket, I take out a map for lower Kakombe valley and start to work. I need to locate all the palm groves in the park; I want to find out where food is most abundant for the chimps. Lower Kakombe is the heartland for the Flo family; Flo died near the streambed just down there. Her daughter Fifi started to transfer into another community when she reached adolescence, but changed her mind and eventually inherited Flo's core area. The lower Kakombe seems to have more palms than most other valleys. Maybe that's why Fifi, like her mother, has been able to raise so many children.

Now to survey the higher valley from the knoll where Jane brought her telescope each day. A shy troop of baboons barks in alarm and retreats into the forest below. There are just a few palm trees up this tributary, none at all in that one over there. The clouds close in, and a shaft of sun spotlights a grove of trees atop a streaked grey cliff. Down below, a white-backed palm vulture soars over the forest floor.

I fill in more of the map, then carry on higher and higher, through back-lit tussocks of glowing green grass. Up into high spindly trees delicately flocked with eiderdown leaves, their trunks blue with lichen, their branches hung with Spanish moss. Mushrooms lie scattered every few yards along the ground. The high slopes are strewn with boulders of grey and white; pink orchids cling to every nook and cranny.

At the crest of the next rise, the outside world hits me in the face. Steep bare fields have been stuck like stamps on the far side of a ravine just beyond the park boundary, four valleys to the north. The countryside is pale green all the way to Burundi and beyond. The forest has been replaced by fields and low scrub. Metal roofs glisten in the sun.

I march along an open grassy slope above the precipitous face of the highest peak, then through alpine meadows sprinkled with yet more orchids, lilies, and protea bushes. Clouds race in at my feet but nothing blocks the sun, the clear, crisp sun that colors the world with an exhilarating intensity. I reach the summit in a swirl of butterflies.

But to the east, just outside the park, the rugged terrain has been covered with fields and small villages. The cultivated slopes are too steep to hold the soil after a heavy rain. Those mountainsides were forested the last time I came up here, but the population of Tanzania has doubled since then, and thousands of refugees have moved in from Burundi and Zaire.

To the south is the Kigoma airstrip and the fields we flew over on our way in. Westward, across the lake, the Zairian escarpment has also been checkered with fields. Fields, fields, and more fields in every direction. The forested slopes of Gombe stand out as a lone jewel set in an endless band of settlement. An urban park. Twelve square miles.

I pull the Global Positioning System instrument out of its black bag, and within seconds it has picked up two, four, then seven satellites. The dial shows my current latitude, longitude, and altitude. Every few seconds the coordinates change by a few thousandths of a degree, the elevation bobs up and down by a few hundred feet.

Clouds silently fill the lower valleys of the park; rain falls on the lake. But away to the east, I can hear the sounds of children laughing and playing, women chopping wood, a Land Rover grinding up a distant hill.

After an hour, the machine has revealed my location to within a few yards: 4°40.962′ south; 29°38.701′ east, my altitude is 5240 feet.

I have left the forested valleys to measure the precise locations of a number of landmarks around the edges of the park. Cartographers in Minnesota will use these as reference points to create a new map from a series of aerial photographs.

The GPS scans the skies for military satellites, but the Pentagon scrambles their signals to reduce the machine's accuracy to within a hundred yards—only American soldiers are permitted to determine their whereabouts to within a few inches. The scrambled signal makes it look as if you are wandering around drunkenly, but since it never misleads you by more than a hundred yards, you only have to wait until it staggers around for a hundred yards in all directions. The Pentagon doesn't mind if you eventually work out your precise location, so long as you aren't moving with the speed of a guided missile.

Zipping the GPS into its bag, I continue through the high country, following a path that zig-zags down a cliff-face and winds outside the eastern boundary of the park. Two teenage girls approach along the trail, each wrapped in matching squares of brightly patterned cloth. They cover the uneven ground with regal grace, walking erect with their colorful parcels on their heads. The effect is straight out of Gauguin, but they turn and flee in giggling terror as soon as they see me.

Descending into a broad open valley, I pause by a fork in the path where a middle-aged man halfheartedly hoes his field. His corn is choked with ferns, and the ground is sticky and wet. He is Waha, and pleased to be greeted in Kiha rather than Swahili. I ask how to get back to Kasekela; he points along the trail that passes a couple of flimsy huts with mud-covered walls and loosely thatched roofs.

Paths split off to the left and right. I engage more pedestrians with more Kiha greetings and ask for more directions. Someone asks in English if I'm hungry; someone else stands still with his hand out, demanding a gift. An old man plods slowly along, carrying a spade and a spear. The fields grow thin crops of cassava, corn, bananas, and pineapples. Storm clouds have gathered overhead; the slightest breeze rattles the banana leaves and sounds like rain.

A woman pounds cassava in a steep-sided wooden bowl. Her toddlers swarm around her feet; the smallest sits placidly on an open square of cloth. She is only in her mid-twenties, but already she has had five or seven children and lost half of them to malaria or dysentery or flu. She is still young and pretty, but she knows altogether too much about life and death.

Continuing my circuit, I enter the lakeside village of Mwamgongo at the northern border of the park. The population here must be close to a thousand; most houses have tin roofs and cement-block walls. People sit outside in the shade of the mango trees. A pair of women weave each other's hair; one woman sits on the bare dirt and leans her head blissfully against her hairdresser's knee. Standing in a loose circle, a dozen men have gathered to watch a game of *bao*. The alley echoes with the triumphant rattle of round black beans being flung in each pot during a long complex move. A kid pedals by on his bicycle, the only wheeled vehicle in town. It has no tires, just rags tied around the rims.

The fishermen have all come home for the full moon, and the beach is lined with catamarans and dugout canoes. I sit on a rock by the lakeshore near the northwest corner of the park. The GPS spits out its random dance of misinformation. I write down the coordinates and wait for the pattern to complete itself.

A crowd of children gathers around, curious about the crazy *mzungu* and his mysterious black box.

"What is that? A radio? A camera? What are you doing?"

I sit flustered for a minute, constructing an answer in my limited Swahili.

"This box receives messages from the sky, but they are all lies. I must wait here and write down all the lies until I can finally figure out the truth."

They move in closer. "What's that?"

"Where am I."

FRIDAY, 22 NOVEMBER

This morning I have Sibelius's Fifth Symphony in my head. The sun is shining across the high rugged slopes above us. A network of narrow terraces crisscrosses the slopes, casting a delicate lattice of slender black shadows. Sibelius's cascading ostinati reflect the repeated elements of the landscape and lift me higher and higher. Anthony stops to take a picture, but his camera jams. This is one of those views that never works on film anyway; nothing captures the glowing brilliance of the color or the full sweep of the countryside.

We carry on up the Mkenke *watu* (people) path, then leave the trail to reach an overlook of Kahama valley. Kahama is where the southern chimp community once lived; over there are the cliffs where Ruth fell. Down on the lake, close to shore, a shabby white steamship carries passengers up to Mwamgongo and the Burundi border.

Kahama used to be the southernmost valley in the Kasekela community's range, but last night the Kasekela males nested two valleys further to the south. Has something happened to the Kalande males?

Gombe is slightly wedge-shaped, and the Kalandes live at the narrow southern tip of the park. All those people pressed up against the park boundaries are never more than a half-mile away. Large numbers walk through Gombe each day on these *watu* paths. Farmers come from the eastern side of the rift to trade vegetables for fish; long-distance travelers seek out the level paths along the lakefront.

These are not the healthiest people in the world. This climate breeds disease, everyone lives in poverty, hygiene is appalling, health care doesn't even bear thinking about. Travelers drop the odd scrap of food, the fishermen abandon old clothes in their huts. Every now and then, when the fishermen have left the park, the chimps go down to the beach and enter the abandoned huts. They suck on old pieces of cloth, chew half-eaten stalks of sugarcane, and eat the ashes in the fireplaces. Chimps can be infected by the same diseases as humans, are vulnerable to the same epidemics.

The Gombe chimps were devastated by polio in the '60s and flu in the late '80s. Each outbreak killed a quarter of the Kasekela community. Perhaps the Kalandes were hit even harder by the flu; perhaps the next epidemic will wipe out a whole community. As a population shrinks in size, the animals are forced to inbreed and become even more susceptible to disease.

And disease is not the only threat. There, lying in the grass beside the path, is a piece of wood carved into the shape of a slingshot. It is the trigger of a snare, just like the ones we saw in Adam Hill's office at TGT.

In the old days, we only worried about two things while wandering through the forest. First and foremost were the snakes. There are more poisonous snakes at Gombe than anywhere else in East Africa: black mambas, green mambas, puff adders, boomslangs, forest cobras, spitting cobras, Egyptian cobras, and water cobras. You learn to watch your step. Second, we worried about the small herds of Cape buffalo that used to roam the park. I say "used to roam" because the buffalo are all gone now, eliminated by poaching. There used to be lions and hyenas in this part of the country, but they, too, are long gone.

What is next? Whole villages of Zairian refugees have settled just outside the southern end of the park. They come from tribes that eat monkeys. The baboon troops around the research center are doing all right, but what about the baboons in the rest of the park? Some tribes eat chimps, too.

The Serengeti is so large that most of the surrounding population might conceivably profit from the enlightened management of its wildlife some day. But Gombe is so very small, and surrounded by so many people.

Jane provides some of the few salaries in the area, but not everyone has a stake in Gombe's future.

Anthony and I walk beneath the dense canopy on the valley floor. A troop of red colobus monkeys crashes through the treetops above us, chirruping in alarm and leaping from branch to branch. We cross a narrow streambed littered with pale, square boulders and enter a large palm grove. Potsherds litter the ground, a reminder that before Gombe became a reserve, there were grass houses in all these valleys. A small number of people had always lived here.

Do any chimps still survive in the remaining patches of forest outside the park? "It's hard to say," Anthony replies. "If you go in looking for chimps, you have to be careful not to be too obvious about it. If the local people suspect you've come to survey the area, there's a danger they might try to kill all the chimps. They're afraid that the land might become a chimp reserve and that they'll all be evicted."

Apolly sits beneath a framed certificate from Jane commemorating his twenty years of service. The chimneys of his hurricane lamps are half-covered with soot, the room is filled with the odor of burning kerosene. Hanging on the wall above his narrow dining table is a tattered bulletin board. It is covered with faded black-and-white photographs taken in the years before the kidnapping: Emilie waves from a canoe, Anthony stands among the baboons, students pose like Tarzan. Pinned between the photos are pictures cut out from magazines: a classroom teacher, a Hindu holy man, Saddam Hussein, a tourist ad for Burundi. In the middle of everything is a color snapshot of my daughter Catherine, taken when she was only a few months old.

I ask Apolly if everyone will be willing to work overtime for Christine this weekend. Apolly takes off his black-rimmed glasses; he has a headache, possibly a bout of malaria, and he may not be able to go out tomorrow. But the rest of the baboon followers are eager to continue. Trailing a baboon around with a tongue depressor may not seem very glamorous, but the shit-collecting campaign has been one of their moments of glory.

Outside, moonlight bathes the alleyways of the staff camp. All the doors are open. Everyone has tuned in to the same radio station, and the music dances from one end of the village to the other. Flickering candles and dull hurricane lamps shine through the open doors. Bofu, the old handyman, sits in his bare room on a hardwood stool beside an open wood fire and a blackened aluminum pot. He stares silently at the coals and waits for his dinner to cook. With his gnarled face and white *balagashia* cross-lit by the low flames, he presents a portrait worthy of Rembrandt.

Fires smolder in most houses, the smoke pungent as incense. A young boy stands outside one door and watches the *mzungu* stride by; a friendly female voice cries "*Karibu!* (Welcome!)" Quiet conversations flow out of the houses and carry the gentle sounds of companionship. It is a world of family and friendship of another time—a time in which people amused each other with tales of their own adventures, through the quality of their own talents.

As I walk back to the forest, the music and conversation are supplanted by the chirps of crickets, the clanging bleeps of hammer-headed bats, the croaks of tree frogs. I pass into jungle as timeless as the full moon itself, all fat and wet and shining through a diaphanous haze that clings softly to the warm curves of the earth.

SATURDAY, 23 NOVEMBER

The sky is heavy and black, the lake has turned slate grey. A storm is gathering fast at the top of the rift. Over in Zaire, the morning sun still shines on the pale green mountains, the last trace of color in the world. Anthony and I climb inland at the mouth of Kasekela stream and follow a path into the dark, dank forest. After a quarter mile we gaze down upon the lower valley and map the palm trees in the dim light. To the south, thunder is rumbling constantly. God has indigestion this morning.

Above us the heavens have gone mad. Thick, tubular clouds streak sideways across the sky, unrolling like immense carpets. One band veers out over the lake, then starts to swing back toward us like the arm of a hurricane. Another massive arm sweeps out from the rift in the opposite direction. They are going to collide straight above us.

The thunder has become more urgent, and trees tremble in the uncertain air. Zaire is in shadow now, too, all color gone. We descend to the valley floor and cross the dry streambed just above the course of the empty

Kasekela waterfall. A baboon troop is perched on top of the cliffs. Anthony points out some of the animals I used to know. They pause on the rocks to groom, then continue up the valley. We follow them to a large palm grove.

A few large raindrops are starting to fall, and the thunder has become a constant roar. The massive clouds bend and swirl as if all the forces in the world were concentrated above our heads. We can feel the weight of all that water suspended in the air. The ground has started to shake from the thunder. The storm finally breaks free, and we can only brace ourselves, there is no place to hide. The air is white with rain; we stand facing the ground to keep from choking. The baboons sit like Buddhas with their eyes closed.

My anorak shields me from the pounding torrent, but Anthony is wearing only shorts and a cotton shirt. He looks miserable. The cool rain keeps beating down, draining him of all heat. He has to dance around to get warm. The water rushes down the slopes past our feet, washing the hillside, feeding the swollen streams.

After half an hour, the worst of the storm seems to have passed. There is no more thunder, and the baboons have shaken themselves off and resumed feeding in the palms. We pull ourselves up to the top of a small knoll where we can hear from its violent roaring that the Kasekela waterfall has been brought to life. Anthony is keen to carry on but he is shivering uncontrollably; there is no point staying out any longer.

Slipping and sliding down a hillside slick as butter, we stumble onto the sodden beach. The storm has stirred up the lake, and rapid, driving surf pounds the shoreline. Wet and shiny now, the stones reveal patterns of chaotic uplift and shattering collapse. White rock is injected with black streaks, green stones infused with red. Gneiss, schist, agate, jet, and jade have all been crushed and splintered, their pottery cracks mended by crystalline seams of quartz; the passive tokens of large-scale forces too diffuse to measure in the course of our own lives.

After lunch, Anthony collapses in bed; his chill seems to have brought on a bout of malaria. I spend the dripping afternoon in one of Jane's closets looking for data that were missing from her house in Dar. In among old notes and files are dozens of fan letters from all over the world. In broken

English, in careful cursive scrawls, on ancient typewriters, people from every continent have written to tell Jane of their great admiration for her and her work. People from all walks of life have been moved by her books and her films; they ask for her autograph, for advice, for information. As long as I have known her, Jane has always answered each and every letter.

Jane's contribution goes far beyond her discoveries about the chimpanzees. She has almost single-handedly changed the way humans view other species. In the early days, she was scorned by the scientific community for endowing chimps with human qualities and emotions. Most psychologists and behaviorists viewed animals as automata that were only capable of reflex responses to specific stimuli. But chimpanzees are no mindless machines; they plan ahead and anticipate the consequences of their own behavior. Jane's findings could not be ignored for long. Now, almost everyone acknowledges that animals do indeed have their own desires and goals, and that they lead the most intricate and engaging lives. Putting yourself into the skin of another species is often the best way to understand it.

But we lucky few students gained something else from working with Jane; she made us feel part of a noble enterprise. We were young and starry-eyed, raw clay eager for guiding hands, and most of us left Gombe with a strong sense of the value of community, of working together on the hard tasks—what Jane always called the "Gombe spirit." Gombe has seen its share of ructions and schisms among the students and staff, of course, but research here has flourished precisely because Jane had a vision that could not be realized by one person alone.

Anthony reemerges in the late afternoon, pale and exhausted. One of the chimp followers, Msafiri, comes in to talk, and they sit down at a low table to drink tea. Anthony's Swahili is far better than mine, so I can only occasionally follow the conversation. They discuss Msafiri's family, the state of his home village, affairs of the staff camp, and the chimps. The Kasekela males have come back north again after their excursion into the deep south. Yesterday, Wilkie, the dominant male, was challenged by one of the younger males, but the young challenger has no reliable ally and cannot stage a coup without help.

Anthony first came out from England in 1972 as an administrator and an assistant on the baboon study. He too became enthralled with the baboons and enrolled as a graduate student. His thesis research was terminated by the kidnapping, and he spent the following year in central Tanzania studying another population of baboons.

By 1975 the Tanzanians had become so familiar with the research routine that they carried on perfectly well without us. Foreigners were no longer allowed to work at Gombe, but Jane talked to Ettha and the followers over the radio each day from Dar. After a few months, Jane herself was permitted to return, first for a week or so at a time, then for four to six weeks. Other foreign scientists started to filter back during the late '70s, and Anthony returned in the mid-'80s.

Now Anthony spends six months each year at Gombe. His own research focuses on the park's vegetation and the long-term histories of the baboons, and he is vital to maintaining the Gombe spirit. The staff adore him. He always has time for their troubles and works tirelessly for their well-being—he even helps them with their income taxes.

Msafiri, Hilali, Eslom, and the other chimp followers have suffered through a hundred storms as violent as the one that hit us today, but they still worry about the health of the older chimps, root for favorite males in dominance struggles, speculate how far the Kasekela males will penetrate the strongholds of the other two communities. Every evening Anthony goes over to the staff village to discuss who they plan to follow the next day, but he has to work hard to convince them to watch the females. We are just as interested in the feeding and ranging of female chimps as in the exploits of the males, but the followers don't like following females—it is too easy, too boring. They want to go with the males on a heroic adventure to some far-distant valley—the harder the follow, the better.

The followers here want to work harder than we would like; they want to take risks. These are not laborers but explorers caught up in the excitement of discovery. What is going to happen next? What will we find? The work is physically and intellectually demanding of everyone involved in the research. A team spirit pervades the whole enterprise.

After moving to the Serengeti, Anne and I found it hard to accept the Gombe spirit as unique. We succeeded in inspiring the Mweka students to listen for roars for a few days. But the student who fabricated his data was all too typical. He *said* he wanted to do research, but it was just a ploy to get a car. The car figured in his dreams in the same way as Kisiri's long-sought bicycle. The real motivation was business; the research was merely a means to his own end.

But the research at Gombe has become a part of the local culture. The staff camp is a local village, the work is a tradition. The research has encouraged these people to do the most unlikely things, and they do it together.

SUNDAY, 24 NOVEMBER

Narrow shafts of sunlight brighten the understory, and the ground is still wet and slick. Trail cutters have hacked out a system of broad footpaths through the park, but innumerable animal paths branch off in every direction. Somewhere I've made a wrong turn, and the narrow, arched tunnel in the undergrowth has become a dense tangle of branches and vines.

Emerging into the open sunshine, I clamber up the slippery ridge top, occasionally grabbing at a few long blades of grass—literally clutching at straws—to keep my balance. Atop a high knoll, I outline the extent of the palm groves on a rough map of Rutanga valley and begin to take readings with the GPS.

Along the valley slopes, clouds of lace-winged insects rise like wisps of smoke, attracting dozens of hawks, kites, and eagles that take turns swooping down to feed. When the ground has been softened by the rain, termite colonies launch thousands of winged reproductives. Flying termites are loaded with enough energy to dig a new home and to spawn a family of millions. Fried lightly in oil and sprinkled with salt, they taste like exotic nuts.

I've seen enough up here today; I need to go help finish the baboon work. Besides, the sight of food has made me hungry . . . Now where was that path? The forest edge all looks the same; the passage was masked by a wall of vines . . . Is this it?

Leaving the sunshine behind, I follow a vague line downhill for fifty yards, then reach a dead end. A second slippery path peters out after a hundred yards. My feet fly out from under me, entangling myself in the thicket. I have no choice but to keep going down.

I crawl on the ground face first, squeezing between vines, lifting branches out of the way, then sliding uncontrollably down the bare soil until a dead branch snags my shirt. A bug tunnels its way under the strap of my binoculars. The metallic chirping of a cricket sets the Grateful Dead going around and around in my head; the electric guitar noodles insistently through "Dark Star:" Rawp-DAW, daw-duh daw-da-da-daw. Rawp-DAW, daw-duh daw-DAW-da-daw. Rawp . . .

The ground becomes even steeper, I am almost skiing, but skiing through barbed wire, reaching from one sapling to the next, brachiating along the ground. A thorny vine grabs my ankle and engraves a bracelet of blood across the top of my foot. The ground levels off slightly until I reach a rocky ledge. The top of a cliff perhaps? It might be wise to go around.

Bup ba-dup bup bup ba-duh-duh duh-duh duh.

Don't tell me I have to go down through there. After straddling branches and sliding hard into a briar patch, my face is down to the ground, and the earth smells of methane. The ground is alive, and it stinks.

It's hot and steamy, and I'm wet, filthy, scraped, scratched, and starving. I've just about had it. Sure, it's warm here today. The sun is shining out there somewhere, and this is an exotic part of the world. I enjoyed doing this sort of thing in my youth, but it's time to let the next generation spend their own time scrambling about, tearing themselves up, and being poked in the ear by, phooey, a rotten piece of vine. A clean, dry computer never gave anyone hookworm or schistosomiasis or god knows what else is on these sticky slopes.

Yikes, that's steep! . . . Don't slip on—damn! . . . Okay, get up, swing over that gorge, hop down onto that flat rock. Not too bad—I've reached the bed of a small stream that forms a narrow canyon. But then, in mid-leap between two rocks, a low, looping vine grabs me around the neck. I dangle on one leg for a moment, trying not to lose my balance and hang myself.

Deedle-deedle deedle-deedle dee.

Sudden death averted, I carry on down the rocky ravine, then stumble onto a broad, clear trail. Striding rapidly along the freeway, I remind myself to watch out for snakes. The high-vaulted forest surrounds a large, flowing stream, its rapids glistening in the dappled green light. I scale down a low cliff and finally burst out onto the open beach as the music comes to an end:

Ladyfinger dipped in moonlight
Writing "What for?" across the morning sky
Sunlight splatters dawn with answers
Darkness shrugs and bids the day good-bye.

All along the beach, dozens of flying termites have landed and shed their wings. Hopeful bachelors broadcast their charms by placing their heads to the ground and their tails in the air, emitting sex pheromones. Newly mated pairs are already scurrying along in tandem, ready to start housekeeping. But they have all landed on the barren beach where no colony could possibly survive, far from food and shade. I have brought along a small jar, and after collecting newlyweds for a mile or so along the beach I am ready to start cooking.

But now I'm kicking myself. I should have collected each couple separately and preserved them for genetic assay. Termites are among the most highly cooperative animals in nature. They are so cooperative, in fact, that several biologists suspect them of being highly inbred. If termites habitually inbreed, members of the same colony would be extremely close kin, close enough kin for workers to forego breeding and to act as sterile helpers to the king and queen.

Intense inbreeding may help to explain the comparable social system of a peculiar creature called the naked mole-rat. Naked mole-rats, NMRs for short, are the mammalian equivalent of termites. They live in large underground colonies in the deserts of eastern and southern Africa. Blind, ugly, and incapable of controlling their body temperature, they are frequently described as penises with teeth. Each colony contains a breeding pair that viciously prevents any subordinate from reproducing.

Colonies are so widely scattered, and NMRs are so ill-prepared for above-ground travel, that a young NMR can never move from one colony to another. Consequently, they spend their entire lives in their colony of birth, and a fortunate few inherit the mating duties of their like-sexed parents. But their spouse will inevitably be either their opposite-sexed parent or a sibling. Subordinate NMRs help maintain the tunnel system, actively protect the colony from snakes, and bring back food for the babies. Given such a high degree of inbreeding, those babies are almost genetically identical to the helpers. From a biological point of view, it hardly matters who breeds within each colony.

Termites are not imprisoned in their natal colonies like NMRs. The orgiasts wriggling around in my jar may have traveled hundreds of yards and originated from several different mounds. Cooperation can also evolve without intense inbreeding. Lions and chimps, for example, cooperate with their siblings and cousins. But the kinship patterns of termites are not well known, and we could easily measure their family ties with the same genetic assays we use on the lions. If both members of a mated pair flew out of the same colony, their genetic fingerprints would be very similar.

I have lived in Africa for years, watching some of the most conspicuous and charismatic animals on the planet. But I don't see baboons, chimpanzees, and lions as pretty pictures in a glossy book. I view them as abstractions, like well-defined characters in a good novel. The social evolution of termites is just as intriguing as that of the largest mammal.

I sometimes wonder aloud if I still need to come back out here each year, but Africa is positively bursting with life. Creatures come at you out of nowhere, posing fascinating questions, demanding to be studied. I may be tired, filthy, thirsty, hungry, bleeding, and sore. The clouds may have built up again from nothing, promising more rain. But I've brought back lunch and an idea for next year. Not such a bad day after all.

Apolly has gathered his staff in the baboon office. Scattered around the room are three hundred plastic vials, each labeled with a date and the name of a baboon. Next to the window, tiny white worms are wriggling around in a dozen petri dishes half-filled with crushed charcoal. Anthony translates to the baboon followers as Christine describes the samples she wants collected after our departure tomorrow.

Anthony and I go over to the attendance sheets pinned to the bulletin board. The baboon followers fill in their daily census of each troop and record the reproductive state of each female. This status is easy to see because a female baboon wears her heart on her buttocks. The bare skin starts to swell a few days after menstruation and reaches a maximum as she approaches ovulation. If she conceives, her bottom remains flat but turns bright vermilion. After she gives birth, the skin fades to grey.

We go over each name on the attendance sheets with Apolly and the followers, making sure we know the correct initials, birthdate, and mother of every infant and the troop of origin for each of the males. The attendance data give us the family structure of each troop and the reproductive performance of each animal. Multiplication is the name of the game, and the winners survive to be counted.

Work completed, we congratulate the baboon followers for a job well done, pay them their overtime, and spend an hour shaking hands and saying good-bye. The followers hurry back home as the rain starts to fall, and Anthony goes to his office to prepare for the trip to Kigoma. Christine and I spend another hour packing each vial so that it won't tip over inside the large cardboard box. The rain finally stops, and Christine wipes the office shelves with disinfectant.

After hauling the heavy box of samples back to the Cage, I go for a final swim in the lake. The sun has set over the mountains of Zaire, and the dull red sky is as flat and faded as an old photograph. I float peacefully in the warm water, far from shore, totally absorbed in the fading outlines of the hills and valleys above.

I had quite forgotten the excitement I used to feel for this place, the landscape, the animals, the colors, the smells. There is a sheer joy in working someplace that you've come to know inch by hard-won inch. Keeping up with a chimp or a baboon on these steep slopes feels like a minor miracle. Scratches and scrapes are battle scars to be worn with pride.

But there is no getting around how unhealthy it is here. Everyone has malaria; most of the staff are infected with worms. In the months before the kidnapping, I suffered from chronic dysentery as well as the well-timed case of hookworm. How I hated those urgent trips to the loo in the old days. Getting up on hot, rainy nights, trudging outside, and sitting on a steaming wet toilet seat, being tickled by the cockroaches that congregated just below the lid.

Ten days at Gombe feels just right, thank you. I'm ready to quit while I'm ahead.

MONDAY, 25 NOVEMBER

At 5:30 AM the world is still awash in moonlight. We finish breakfast at Jane's house, and Anthony goes outside to prepare the boat. Then, "Oh, no."

A cramp. Not just any old cramp, but a Gombe cramp. Me and my big mouth. I was so thirsty yesterday, I must have drunk gallons; gallons that washed into the stream after the rains on Saturday. Nothing tastes quite like the bathwater from your local baboon. I run to the outhouse and make it just in time. Well, at least I'll be empty between here and Kigoma. We step into the open wooden boat, and the followers congregate for final good-byes.

"*Kwa heri!* (To happiness!)"

We stop by the ranger post at the southern end of the park to pick up one of the ranger's wives. She wears a low-cut blouse with fluorescent green stripes and a skirt printed with turquoise and chartreuse squares. She nurses a small baby boy, covering herself casually with a black cotton shawl, itself a riot of blue roses and red daisies with chocolate-brown leaves. The shawl carries a large Swahili slogan that translates roughly as "I don't see any problem with the way I am." The boat rises and falls as we pass over long, slow rollers. The lake surface breaks the reflection of sky and hills into broad, colorful brush strokes, an abstract African pattern.

Outside the park boundary, the shore is lined with permanent villages and fleets of outrigger canoes. The forest is almost gone. The slopes have been plastered with bare brown fields too steep to climb, let alone cultivate. Older fields have been abandoned, some have been reclaimed by a pale green scrub, but rain has carved others into great gouges down the hillside.

After an hour, we pass a small beach just before Kigoma harbor, the beach where Barb was released a week after the kidnapping. During the night, soldiers paddled across the lake and dumped her to deliver a set of demands.

The students had been kidnapped for political purposes. They were being held by a group called the People's Revolutionary Army, PRP, a Marxist organization committed to seceding from Zaire and forming an independent state. They controlled most of the province across the lake from Gombe, and their strongholds included the small villages along the escarpment. Barb told us that the Zairian security forces sent gunboats up and down the lake to fire into the forest each day. As the bullets whizzed by overhead, the students were forced to hide in trenches beside the rebels.

One of the most unforgettable passages in *Heart of Darkness* is Conrad's description of European gunboats keeping order in the West African colonies, sailing up and down the coast, firing cannons blindly into the jungle. Waging war on a continent.

Only the masters have changed.

November 25th

Dear Anne,

The plane is late. We left Gombe at six this morning in case Air Tanzania arrived early again, but our flight has been delayed till late this afternoon. We have been assured that "the aircraft has not been detained by technical difficulties."

The main street of Kigoma is lined with the same set of shops, the road is still muddy and full of potholes. Some of the market stalls sell nothing but empty bottles, others sell secondhand clothes donated by Western charities. But the place is crammed with goods, and the local cinema is showing *In Gold We Trust.* The alleys are crowded with a throng of surprisingly well-dressed people from all around the lake. The women wear a terrific variety of cloth from Zaire, Burundi, and Tanzania, and their hairstyles show a real flair that I had quite forgotten: sputnik-spikes, spiderlike loops, tightly tucked braids that mimic old-fashioned perms. All those flamboyantly colored clothes parading the street bring the muddy sidewalks to life with a festive atmosphere that is quite unlike my memory of Kigoma as a casualty of socialism.

I am waiting in the restaurant of the Lake View Hotel, unable to resist a doughnut even though my intestines are in chaos from drinking too much Gombe water. After nine days of inadequate diet combined with so many days climbing the upper valleys, I am unable to fast.

The latest news from around Africa is rather chaotic as well. Serious tribal fighting has broken out in the capital of Burundi. There is new fighting in Rwanda, too. The Rwandan army has bombed Ugandan villages for sheltering Rwandan rebels. And don't forget Zaire. Hundreds of people have fled in the latest attempt to overthrow President Mobutu. At the other end of the lake, cholera is spreading from Zambia into southern Tanzania. Oh yeah, there was also a big rally in Kenya on Saturday, demanding political reform.

We have moved our wait to the Kigoma airport. The outside of the terminal has been painted bright green and blue. The only aircraft in sight is an abandoned missionary plane that fled from the latest wave of unrest in Zaire.

Bedtime in Kigoma. At about four, we learned that the Air Tanzania plane had taken off from Tabora, but the pilot was sleepy and wanted to rest in Kigoma overnight. We rushed back to town because Anthony had been warned that the hotels would soon be fully booked. Sure enough, the first two hotels were full. We drove to a third, and the taxi ran out of gas. Phileas Fogg, here, got out, asked for directions, and flagged a second taxi. Rest house number three, the Milimani Mess, was also full, but the desk clerk was very helpful and called number four, the Milimani Fisheries Lodge.

Rooms were found at last, complete with sink and flush toilet in the bathroom, but there was nothing in the pipes—water must be fetched from a rainbarrel out in the courtyard. A large sign in each room lists the conditions of occupancy. Item number seven: "No form of chaos is permitted in our environment." Thank heaven for that.

We had dinner at the New Railway Hotel with Mrs. Kawanaka from the Mahale Mountains chimp project. She was surprised by the remnants of my Japanese, although it was easiest to communicate in Swahili. She is an artist, a cousin of Professor Itani. For twelve years she saved up to come back to Tanzania to teach the local women how to dye cloth and mats using extracts from the native plants. She is very warmhearted, even effusive after a full bottle of Safari lager. She decided to leave Mahale three days early, she said, because the longer she stayed, the harder she would have cried while saying farewell.

Love etc.,
Craig

KIGOMA / TUESDAY, 26 NOVEMBER

A muezzin recites his prayers just down the hill from the Fisheries Lodge. It is 5:00 AM and the piercing loudspeaker projects his whining singsong to every corner of the planet. Between bouts, he exhorts us all to live the life of a true believer, reminds us of our duty to Allah, and generally does his best to keep us all awake. As his final cry echoes around town, every dog in Kigoma joins in.

Our taxi to the airport arrives, and we carefully load Christine's box of fecal samples into the back. Anthony and Christine feast on bananas and mango, but my guts are screaming, so I can only sit and watch. The sky is heavily overcast, the sun must have risen by now but it is still dark. We pass a police barricade next to the Kigoma market and head up the hill through a sudden cloudburst. The windscreen wipers don't work, and the windows won't shut.

At the airport, we hand over our tickets and place our luggage directly onto a rusty yellow wagon, instructing the baggage handlers to keep the box of samples upright at all times. Fellow passengers have gathered on the runway to wait for the pilot, the Asian families keep to themselves in their cars and minibuses. The sky is still cloudy and dark; there is a steady rumbling over the Malagarasi.

Dr. Hans Rosling is leaving today too, and he tells us the latest news from around Africa. Fighting in Burundi has intensified, and a domestic flight in Ethiopia has been hijacked by the Tigre Liberation Front. Dr. Rosling travels throughout Central Africa each year; he has a professional interest in political instability. The abandoned missionary plane from Zaire reminds me to ask him about the PRP.

"I shouldn't worry about them anymore. They've become very respectable these days. They are in charge of the whole province, unofficially of course."

Gazing at the rift behind Gombe with its barren fields and burgeoning population, I ask what hope he has for this part of the world.

"I will tell you the same thing you'd hear from any aid worker in Africa. The key to controlling population growth is education. Educate the women and straightaway the birthrate would be cut in half. The schools are already in place in Tanzania; they are well equipped and the teachers are excellent.

"What this country needs most urgently is intense development. Give these people some reason to *want* an education. If there's no development, everything will collapse.

"Look around Africa and all you see is disaster. But Tanzania has fantastic potential—almost unique to the whole continent. Every one of these little villages has its own equivalent of a Nyerere, a headman who is loved and respected. The local leaders are the country's greatest resource. Now that the government has decided to switch from socialism to capitalism, they are the ones who are making it work.

"Tanzania may seem like a mess, but at least these people have hope. If you want to see a real mess, go to a country that has been caught up in a civil war for the past twenty years. They have no hope and no one to lead them out."

Our pilot zooms up to the terminal in an official Air Tanzania minibus, looking well rested and eager to return to Dar. In the misting rain a team of four men pushes the luggage wagon to the aircraft. We say our farewells to Anthony, but I will see him again before long. He is flying to Britain for Christmas and then coming to Minneapolis in February.

Our fellow passengers sit patiently in the waiting room on pink-cushioned couches. The walls are pink, too, but the doors and baseboards are enameled bright green, the floor red, and the ceiling canary yellow. The counter of the tiny bar is lined with soda bottles and beer and small plastic bags filled with dried *dagaa*. On the wall is a faded blue poster advertising "Tanzania Beach Holiday."

Half of the crowd is African. Women from Zaire and Burundi wear coordinated costumes of brightly printed cotton. Men from Zambia and Tanzania wear short-sleeved, open-necked jackets called Kaunda suits. Most of the rest are Asian. One woman has a red caste mark on her forehead; her silks are red, gold, and turquoise; her sandals have shiny brass straps. An Asian boy wears a T-shirt that says "Latest Fashion"; a toddler's says "Someone Extra."

We board the plane in a steady drizzle. The pilot starts the engines; the props swing around slowly, then disappear into spinning disks. The plane taxis to the far end of the runway and screams back to the east, straight toward low, cottony clouds rising to a dense black mattress that blots out the sky. We climb rapidly into the gloom and are soon swallowed by the storm. My intestines complain when we hit a patch of turbulence, but the air calms for a few minutes and I start to relax. Christine writes a letter, and I sit with my eyes closed.

But then the sky refuses to hold us anymore. The plane plummets, the propellers have lost their grip. I have no idea how high we are, no idea how long it takes to fall from the sky. Is this the price I must pay for letting my life flash in front of me over the past few weeks?

I think of my family, of course, especially my son Jonathan, who is too young to know me. I think about how fortunate it was that I didn't waste money on new clothes or a new pair of glasses before I left America. I imagine that some day there might be a small marker somewhere in the Malagarasi that reads "He was frugal."

But then, within a few seconds, or a few minutes, the plane catches itself with a sharp jerk. So sharp that the soda bottles in the back of the plane are thrown free of their wooden crate. The seat-belt sign comes back on.

The sun is shining on the ground in Tabora, the weather ahead looks calm and clear, and we are delayed again: the plane has been overbooked. Everyone rearranges belongings, and most parents place their children on their knees. But one last passenger, an elderly Asian woman, is still searching for a seat. She tries to look as charming as she can, and there is one more child who could still be relocated to a parental lap. But "Someone Extra's" father, an Asian businessman from Kigoma, refuses to cooperate. The remaining passengers start grumbling at him, incensed by his behavior and eager to take off for Dar. The old lady turns around, and I finally realize the problem—she wears the veil. She is Muslim, he is a Hindu.

The mother loses patience and sweeps "Someone Extra" onto her lap, the old lady sits down, and the father makes a face. The plane takes off once more, and we read a Tanzanian newspaper. Christine laughs bitterly at an article about AIDS. Bus passengers in the Dodoma region complain about how hard it is to prevent the spread of AIDS in the countryside. Buses frequently have to stop in the middle of nowhere for a half hour at a time, and "there is nothing else to do."

But a second article on AIDS is not even remotely amusing. In Uganda, the best predictor of whether a woman will contract AIDS is the educational level of her husband. Educated African men work in cities and spend long periods away from their families. In some countries, the educated men are dying so rapidly that soon there will be no one left to operate complex machinery, no one to run the country. Single-parent families have become so common in some districts that the schools have had to close. Widows can't even afford to buy pencils; most keep their children at home to help raise younger siblings.

I feel increasingly uncomfortable as we leave the plane in Dar es Salaam. The noontime air is like a sauna. Our luggage finally appears on the carousel. The box of fecal samples comes out wet and on its side. The contents seem to be intact; the outside must still be damp from the morning rain.

At the city center, the Skyway Hotel has two rooms on the fourth floor. As we walk down the corridor, a young African woman opens the door to room 404. Inside, three more women sit on the edge of a narrow bed, dressed for the night. I can hardly concentrate to unlock the door to my own room. My cramps are getting steadily worse. Christine's box reeks rather badly, so I set it out on the balcony and run to the toilet.

On the street, Christine and I split up to complete our separate chores. I am starting to feel dizzy. My eyes take longer than usual to adjust to the bright light. I can cope with the pain, but I need to have my wits about me while walking around town. Damn! I forgot to bring the box in from the balcony. While it is faintly humorous to think of someone stealing it ("Hey, Juma, what did you get?"), I don't think Christine would be very pleased.

But I can't make it back to the hotel right now. I've lost too much fluid these past few days and must get something to drink. I'm only a block away from the Kilimanjaro Hotel and its air-conditioned snack bar. Outside the entrance, a pair of white men greet each other loudly and effusively, but when they see me walking by they suddenly lower their voices. Shifty-eyed, furtive, they don't want to be overheard by anyone who speaks English.

I move slowly out of the fierce sunshine, past the faded selection of tacky souvenirs, into the dimly lit lobby. The power is off, so there is no air conditioning. The old wire-service Teletype stands against the far wall.

After we had collected our data from Gombe, Anne and I caught a plane to Dar and came to the Kilimanjaro to see Professor David Hamburg, who had flown from Stanford as soon as he heard about the kidnapping. Waiting for Dave to come downstairs, we stood glued to the Teletype as the paper slowly inched upward:

"Four students were taken at gunpoint from the Gombe Stream Research Center on 19 May. One of the four has since been released . . ."

Jane Goodall had become a visiting professor at Stanford in 1970. Beginning in 1972, Dave Hamburg sent a steady stream of undergraduate research assistants to Gombe each year. I was in the second batch; Anne enrolled as Dave's graduate student at Stanford shortly thereafter. Dave was our mentor and benefactor. He was devastated by the kidnapping. It had never occurred to anyone that something like this could happen at Gombe. The worst risk anyone had imagined was malaria or snakebite. Dave felt an overwhelming personal responsibility for the safety of the three remaining hostages. He worked night and day to secure their release. Various consulates, international agencies, and governments eventually became involved. Henry Kissinger even stuck his nose into the matter at one stage.

The worst aspect of the situation was that there was no way to talk directly with the PRP. But then a PRP lieutenant named Alfonse arrived unannounced in Dar one day, and negotiations could finally begin.

Dave is one of the most persuasive people I have ever known. A colleague at Stanford once described him as a man who could not only change your mind but convince you that you'd felt that way all along. With a lot of official and unofficial help, Dave eventually won back the students' freedom. Carrie and Emilie were released three weeks after Barb, and Steve was finally set free two months later.

Although they were released unharmed, the students' captivity had been no picnic. Their guards had polished their rifles while telling them, "These are the weapons we will use to execute you." The rebels considered American students to be decadent and uneducated and made them spend all day reading Marxist tracts in French.

There was, however, at least one light moment. Two women held in captivity for long periods eventually need certain items of personal hygiene. Emilie tried to explain what it was they wanted, "Napkins, sanitary napkins." The rebels had no idea what she was talking about. She tried again, and then a light went on. One of the rebels went to another hut and came back carrying a package they had stolen from the supply room at Gombe: party napkins decorated with pictures of Snoopy.

Temporarily refreshed, I hope to make it as far as Jane's house without dirtying the backseat of this man's taxi. I have to pick up one last dog-eared stack of data, the "big charts" that were completed in colored pencil on enormous sheets during the early '60s, and that could never be photocopied. They are too precious to ship by air freight.

At Jane's, there is a message waiting from Anne. The data shipment arrived safely in Minneapolis, but there are a number of additional files that she wants me to search for in Jane's house. I make one last painful tour of the data room, but I can't find anything else.

As evening falls on Dar es Salaam, the city is transformed into Sodom and Gomorrah. Standing on the street corner outside the Skyway Hotel are a half dozen provocatively dressed women with long, cornrowed hair. Some are from rural Tanzania, but others are Ethiopian and Somali. Escaping the refugee camps in their home countries, they have merely exchanged starvation for AIDS. The infection rate among the local prostitutes is well over eighty percent.

In the lobby of the Skyway a German in his late fifties emerges from the elevator cursing a young prostitute in tight blue jeans. They walk into the night together, continuing their mock semblance of a domestic dispute. Upstairs, I pass a middle-aged white couple emerging from room 404 in their sensible, permanent-press clothes. They are blissfully unaware that their domicile had served as an illicit powder room this afternoon. I am reminded of the famous headline in a Kenyan newspaper—"Missionaries approve thirteen new positions"—but I don't think this is what they had in mind.

I bring Christine's box in from the balcony, and the smell is much worse than this morning.

At the mezzanine of the Kilimanjaro Hotel we look for a place to eat dinner. Just outside the Simba Grill is a buffet table. The stuffed lion head on the wall has shed ratty tufts of golden mane onto the trays below. Women are ululating in an adjoining room at what seems to be a private dinner. Public dining is in the Summit Room on the roof.

This is another government-run hotel, so dinner takes forever to arrive. At one end of the restaurant, a series of long tables have been set end to end. It is a banquet in honor of a visiting delegation from Japan. The hosts all wear Kaunda suits, but they seem uneasy, perhaps embarrassed by the awkward service or the bad food.

I'm afraid that the grinding pain of my illness just makes people look uglier, and I can't help but wonder whether these socialist officials provide any better service in their own exalted posts. In Tanzania, a significant proportion of the population works for the government, but an important government official earns less than a good houseboy in an embassy home.

Officials are educated and ambitious and have to maintain the dignity of their position. The government has appointed them to office, and it is up to them to raise their own salaries. Customs, immigration, and police officers all expect to be paid directly by the public for their services. But virtually no one is rewarded for what they do in this system, merely for who they are, for the position that they hold. Ambition is rewarded, but society teeters on the brink of chaos.

The Europeans at the next table are no more reassuring. They have been absolutely scorched by the sun; their faces are smooth yet creased, thirty going on fifty. One of the men pulls out a small box full of jewels. The others look like gun runners, soldiers of fortune, the Mafia on holiday. They have been drinking heavily, their ashtrays are piled with cigarette butts, their voices rise. Then, realizing they might be overheard, they suddenly revert to a whisper.

Dar is not the tourist capital of the world. Many of the Europeans here are vagabonds, adventurers, the flotsam and jetsam of the western world who have come out to live on next to nothing, to play Lord Jim, or to help fan the flames of corruption. To burn themselves to a crisp, smoke their cigarettes, and fondle their fatal whores.

DAR ES SALAAM / WEDNESDAY, 27 NOVEMBER

Feeling worse than ever, I stare blankly at the empty canvas bag I brought out for the big charts. Panic—the charts aren't in the hotel room. I knock on Christine's door, but she doesn't have them either. They must still be at Jane's house. We have no time to go back before the plane departs. Christine suggests writing a letter to Anthony, who can bring them to Britain at Christmas.

I am utterly dismayed, not just by having to leave Dar without the data but by a sense of dissolution. I'm meant to be in a position of authority here; people are depending on me, relying on my judgment. It's all starting to slip away.

After finishing the letter to Anthony, we go downstairs and hail a taxi. While hoisting Christine's box into the back of the car, a dozen samples slip out through a large hole in the wet cardboard. Some of the samples must have leaked when the baggage handlers turned the box on its side. There's no time to find another box now; we have to hurry to the airport.

My ears are ringing. When I close my eyes it seems that the taxi is driving in tight circles. We pass the Top Club billboard (cigarettes for men whose decisions are final) and I almost laugh. Talk about coming in like a lion and going out like a lamb.

In line at the departure desk, we try to put Christine's box into a large plastic bag. As we lift it from the trolley the bottom falls out. Three hundred glistening green vials crash and slither over the floor. A fetid cloud fills the room. Our fellow passengers look as if they had rather seen us drop a bomb.

I open my suitcase, pull out the empty canvas bag, and escape to the far corner of the terminal to pay the airport tax while Christine crams all of the samples into our carry-on luggage.

Stowed safely in the overhead bin above you.

At the New Arusha Hotel it's breakfast time, and I can't imagine driving all the way to Ngorongoro on an empty stomach. But then I have an accident on my third trip to the loo. Oh, god—there's no place to wash, or even change pants. Feeling utterly disgusted with myself, I find a taxi to go fetch the car. Peter Jones had said he would store it at the Burka coffee estate just outside town.

The Suzuki starts well, but the rear differential starts clunking loudly when I turn onto the main road. The noise stops on the highway, but the car dies as soon as I reach the city limits. Stay calm, maybe it's out of gas. After I fill the tank from a jerry can, the engine starts up okay, but dies again while I'm driving into the middle of town.

While I am trying to push the car off the road, a grimy looking man offers to help. He is called a *jua kali* (hot sun) mechanic because he has no shop and works outdoors. He opens the hood, checks the distributor, and scrapes the rotor arm until it gleams. The car fires right up but then the rear end starts clunking again.

My mechanic leads me to a greasy side street and introduces the *jua kali* mechanic who specializes in differentials. For the next two hours, I sit behind the wheel in a daze, trying to sleep. Small snippets of song circle around endlessly in my head. Mostly, I'm stuck on the nursery rhyme that Betty, our African child-minder, used to sing to Catherine in the Serengeti:

Paolo usije kucheza na sisi
Una mikono michafu.

Paul, you can't play with us
Your hands are dirty.

I stagger back to the New Arusha at 1:00 PM and try to call Barbie Allen in Nairobi. The lines are down. Christine and I sit in the hotel snack bar and wonder what to do about my health. Christine had a bug for two days at Gombe, but she managed to kick it. This is my third day, and I am getting worse. We don't have time to go looking for a doctor. Christine rummages through her medicine bag and hands me a bottle of sulfa drugs. Three cold soft drinks and a strong belief in the power of medicine restore my confidence sufficiently to let me load the car and go shopping at the market. But the baskets seem twice as heavy as usual, everything is so far away.

We roar out of town and buzz through Masailand. By 4:30 PM we're off the main highway and bouncing along the rough dirt road toward Mto-wa-Mbu. Then the car suddenly dies. Dies completely.

The orange dust settles around us; the late afternoon is still warm and clear. The engine refuses to start.

There is nothing like intense pain and car trouble in the middle of nowhere to concentrate the mind. How does our distributor look now? Pretty good. What about the fuel line?

Just then a car rolls up, a large Toyota Land Cruiser filled with Africans. The sign on the door says UN/FAO.

"What is the problem?"

And for the next hour, they are all over the car. Maybe the fuel line is blocked. A volunteer sucks on the hose and gets a mouth full of gas. Check the spark plugs, the points, the fuel pump. Disconnect, then reconnect every accessible part. The car starts twice, and we drive another half mile or so. But the engine keeps stalling, and finally the battery dies. Our friends-in-need stick with us, though, trying everything they know.

The lead between the coil and distributor finally snaps in two—the copper wire has corroded to a bright green. That's our problem all right, but we have no way to fix it.

"We will tow you."

They back their Toyota up in front of us, attach a long chain, and off we go, rolling along in the middle of their dust cloud for thirty miles.

They deliver us to the *jua kali* mechanics of Mto-wa-Mbu just as it gets dark. We are effusive in our gratitude, but it is no big deal to them. It is what everyone does out here, after all. If you have a car, you look after other people with cars. We each end up towing someone else as many miles as we've been towed, people we've never met, people we'll never see again. You hardly even think about it; it's the way things are.

The mechanics quickly produce a secondhand Land Rover lead which they whittle down to fit the Suzuki. To revive the battery, though, I have to accompany a mechanic down a dark side street to a brightly painted house. The verandah is adorned by half a dozen young women, conspicuously dressed and professionally friendly. We enter a large, well-lit room decorated with crucifixes and portraits of Jesus. The middle-aged woman in the hallway is disappointed by my indifference to her employees. However, she is perfectly happy to sell me two bottles of battery acid.

The car starts again, and I want to escape the malarial zone and reach the top of the rift. Mto-wa-Mbu isn't named the River of Mosquitoes for nothing. We drive off from the market, but the engine is sputtering by the time we reach the entrance to Manyara Park. The car is barely crawling along, the headlights are growing dimmer. We will never make it up the escarpment.

We roll back down to the Manyara campsites. The accommodations may be spartan but at least they have a shower. My room comes complete with a small packet on the bedside table. The translucent wrapper states the contents to be "one colored condom." Mine appears to be green.

MTO-WA-MBU / THURSDAY, 28 NOVEMBER

The Manyara park rangers have towed us to the garage of a Belgian aid project just outside Mto-wa-Mbu. Noel, the young man in charge, has assigned his African mechanic to work on our car. The flimsy garage walls are made from rotting lumber, the roof is loosely thatched, but the tools are all gleaming—freshly imported from Europe. Our battery is sitting on the ground, hooked up to a charger. The mechanic, wearing a fedora, is busily cleaning the carburetor.

Noel describes the Belgian philosophy of foreign aid; a large share of the money comes from the private sector. Locally, Noel oversees a transport service, a seed-oil refinery, a bean farm, and an irrigation revitalization system—all expected to run at a profit eventually and become self-sustaining. The idea is to develop the country's self-reliance. It all sounds wonderful to me, but given my current situation Noel could tell me he's set fire to my house, and I would just stand here nodding at how bloody marvelous these Belgians are.

Noel has more noble chores to perform elsewhere, and Christine has gone off to send a message to Pam and Sarah. I stay put in the garage, encouraging the mechanic, sitting still and trying not to embarrass myself. The medication isn't working. Maybe I should take something to relieve the symptoms. Here is a bottle labeled Imodium. The name sounds promising; it can't do any harm at this stage.

In the front seat of the car in the heat of the afternoon I am increasingly drowsy until I finally close my eyes. I am overwhelmed by visions of the most beautiful flowers, shimmering and then swirling into a complex pattern of vivid colors. They are really about the most *interesting* thing I've seen in years; totally captivating. Everything seems so perfectly pleasant wherever it is that I am, if "am" is the right word. Imodiated.

But then the mechanic taps on the window and asks me to start the engine. He has reinstalled the battery and it fires up perfectly well—as it has so many times before. We take a test drive, and the engine dies and then starts and then stalls when I take my foot off the accelerator. And now the battery is dead again. We do seem to have a few problems here, don't we?

By 5:00 PM, the mechanic has patched everything together with bits of old Land Rovers and broken-down bicycles, and we are ready to try our luck again on the main road. The car still stalls when we let up on the accelerator, and the battery won't take a charge, but as long as we keep moving, we'll be okay. If we remember to park on a hill, we can always roll start.

We reach the Ngorongoro Conservation Area just before the gate closes at seven o'clock, wind our way up to the rim in the dark, and arrive at the Crater Lodge in time for a bath before supper. Unlike our lodge on the outward trip, this one is privately owned. Guests stay in small log cabins that are clean and in good working order. There is no sign of Pam and Sarah, but the receptionist says they had arrived yesterday. Perhaps they have gone down into the Crater.

I take my second dose of Imodium of the day, and we walk down to the restaurant to discover that today is Thanksgiving in the US. The buffet is loaded down with fruit, cakes, vegetables, and the flesh of dead animals, smothered in colorful sauces. Some time between soup and the main course the Imodium starts to hit again, and a peculiar mixture of pain, drowsiness, and dizziness forces me to close my eyes.

I am immediately oblivious to the world around me. In the darkness, I can see only a narrow gate. Above the gate is a golden sign that reads FUNSD. Is this the misspelled entrance to the next research grant? Or to an exciting town in South Dakota?

I know I should be in bed, or at least avoiding all food, but these are harried days. If I'm not better by the time I've delivered Christine to Pam and Sarah, I'll take a bus to Arusha and fly back to Nairobi, or maybe just go back home to Minnesota. Being a glorified tour guide is no fun at all (not even in South Dakota). If my remaining time is short, I'd rather spend my final days with my children.

I am barely able to climb up the hill after supper, and stop frequently to rest. Three Cape buffalo graze the lawn a few yards off the path but pay no attention to me.

Back in my cabin, I go to bed with a Gideon Bible, reading Revelations. Apocalyptic visions dance before my eyes. Demons and chimeras.

During the night, the Cape buffalo graze so close to the cabin that their horns crash against the wall behind my head.

I dream of a beast with seven heads and ten horns.

Part III

Red Rover

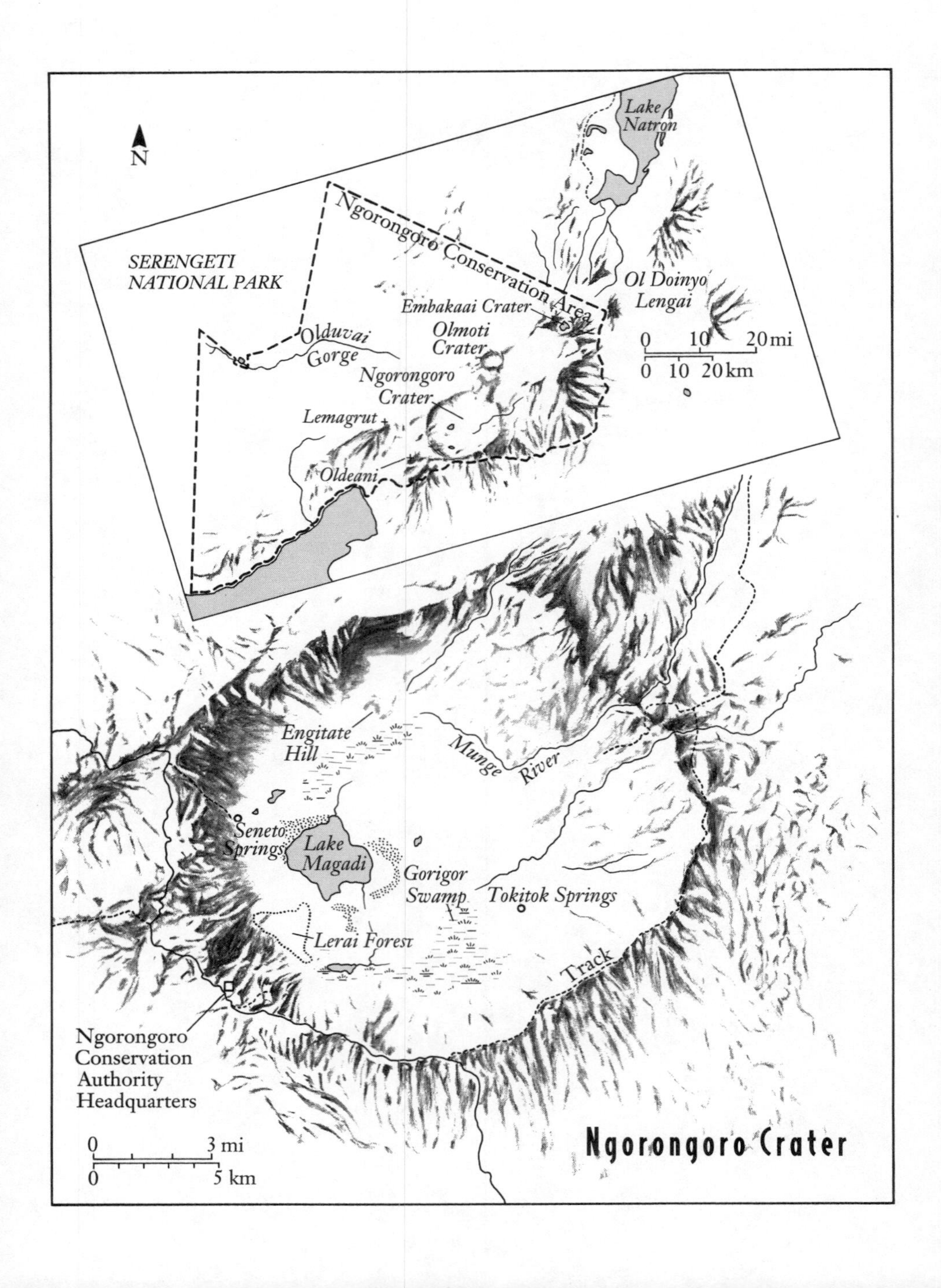

N
Lake Natron
SERENGETI NATIONAL PARK
Ngorongoro Conservation Area
Ol Doinyo Lengai
Embakaai Crater
Olmoti Crater
Olduvai Gorge
Ngorongoro Crater
Lemagrut
Oldeani
0 10 20 mi
0 10 20 km
Engitate Hill
Munge River
Seneto Springs
Lake Magadi
Gorigor Swamp
Tokitok Springs
Lerai Forest
Track
Ngorongoro Conservation Authority Headquarters
0 3 mi
0 5 km
Ngorongoro Crater

NGORONGORO / FRIDAY, 29 NOVEMBER 1991

The sun beats down on the roof of the Lerai Cabin, but in the shade of the verandah I am neither hot nor cold. Minor fluctuations in temperature don't concern me anymore. I am sitting in a stupor in a great circular hole in the ground, ten miles across, with sheer walls that soar vertically above me all the way around. I am at rest in the belly button of the world.

The two-thousand-foot walls of Ngorongoro Crater attest to the hazards of excessive ambition. Two million years ago, a volcano the height of Mount Kilimanjaro exhausted itself while climbing ceaselessly into the sky and collapsed under its own sheer weight. Lesser vents on Ngorongoro's shoulders aimed at more moderate heights and survived to look down upon their fallen rival. From my verandah on the Crater floor, I can see the summits of Olmoti and Embakaai beyond the far wall to the north. Hidden behind the near wall to the south are Lemagrut and Oldeani. At the base of Oldeani a large lake once perched atop the Crater rim, but the retaining wall collapsed and sent piles of chocolate-brown rubble down onto the Crater floor.

The highlands wring moisture from the sky and excess water flows year-round into a shallow alkaline lake at the eye of the Crater. The

Munge River pours down the northern wall from its source in the broken caldera of Olmoti. Springs emerge from the walls to form swamps: Tokitok to the east, Seneto to the southwest, Gorigor to the southeast. On either side of the cabin, a series of smaller springs sustains a small forest of fever trees called Lerai.

The Crater floor is carpeted wall-to-wall with a lush savanna. The rich montane soil and generous seasonal rainfall support more than twenty thousand antelope and their associated predators, the highest concentration of large mammals in Africa. During the rains, dense clouds of insects and innumerable bodies of water attract large flocks of migratory cranes, flamingos, widowbirds, and waterfowl. During the rains, the intensity of life in the Crater can be almost overwhelming.

But the landscape today is yellow and parched. An hour ago, Christine diagnosed my affliction as either giardia or amebic dysentery, handed me a small bottle of Flagyl, and drove off to find some lions and to search for Pam and Sarah. I am propped up on a dusty foam mattress, no longer able to walk. My mind keeps spinning in the same tiny circles. All I can do is lie still and listen to myself think: Should I go back to Arusha to find a doctor or just stay here and expire? Feed myself to the hyenas of the Garden of Eden. Dust into volcanic dust.

When Tanganyika was a German colony, two brothers owned the entire Crater floor and used it as a cattle ranch. Living in paradise, two brothers had a quarrel and divided the Crater in half. One brother built this cabin on a rise above the Lerai forest. The other settled on the opposite side in a farmhouse that now lies in ruins. Two brothers lost their homestead when the British won the war. The stone walls of the Lerai Cabin are two feet thick; the doors and shutters have been locked tight. But from the low-roofed verandah, the view is perfect, the landscape unreal.

Three hours after my first dose of Flagyl and I am already turning the corner. It could just be the rest and the comfortable mattress, I suppose, but this stuff can work quite quickly.

The subsiding chaos in my bowels finally allows sensation to return to the rest of my body. Scratching my back, I find a tick; its head is already well-buried in my skin.

But then, hey, I lead an exciting life.

Dusk is approaching, and a bank of thick clouds has descended across the Crater rim like a great grey lid. Pam and Sarah roll up to the cabin in the red Land Rover, bursting with enthusiasm. They have been having a fantastic time in the Serengeti: seeing lions every day, exploring the park, finding new river crossings, even getting stuck in the mud. The good and the bad are equally exciting—it is all a big lark, an adventure. The lions are brilliant. The car has broken down a few times, but the rangers and mechanics have all been so very helpful. The action never stops!

Christine seems to be stranded down on the Crater floor. We can see the Suzuki from here, and it hasn't moved in several hours. Pam and Sarah hop into their car and head off down the hill, eager to take charge and tow her out before night falls.

Talk about getting a shot in the arm—I have been yanked up out of a fog. The dull, gloomy evening has been transformed into a sparkling new day, full of possibilities, new things to see, new worlds to conquer. I realize now why so many people like to teach. It is an excuse to work with the young.

SATURDAY, 30 NOVEMBER

Piled into the red Land Rover, we reach Seneto Springs in time to find a pride of lions lounging about on a patch of bare soil. The lions are grotesquely fat. Uncomfortable with their own excess, they pant sleepily in the early morning sunshine and shift their great guts around. Suddenly they spring to attention; they can hear cowbells jingling up along the Crater wall. Stretched out along a wide trail three-quarters of the way down from the rim, Masai herdsmen are bringing their cattle to drink from the springs. The warriors carry spears, but the overstuffed lions are too terrified to pose any real threat to them or to their cattle. The lions trot over to the springs and disappear into the reed beds.

After Gombe, the Crater is surprisingly dry and dusty. The rains should have arrived here by now, but the roads are still long white ribbons of loose powder. Most of the grass is brown, grazed right down to the ground. The long-horned Masai cattle are emaciated, and even the wildebeest and zebra are looking sorry for themselves—easy pickings for the lions.

During a normal rainy season, the Crater becomes flooded, and grass continues to grow along the receding margins of the lakes and swamps throughout most of the following dry season. The perennially green grass enables large populations of herbivores to remain on the Crater floor throughout the year, allowing the Crater lions to enjoy a more constant food supply than their Serengeti counterparts. Crater cubs never risk starvation and grow up to be larger and stockier.

Although the Crater floor contains one of the densest lion populations in Africa—a hundred lions in seventy square miles—it supports only a half dozen prides. Each pride defends a particular body of water—a stretch of river, a pond or a spring—and this strict partitioning limits the size of the lion population. Most Crater cubs survive to reach adolescence, but only a lucky few are able to breed successfully. Few cohorts of young males can overcome the large coalitions of resident males. Only a few daughters are recruited into their mothers' prides. The rest leave home to seek their fortune, but there is no refuge on the Crater floor, no space to carve out homes of their own. Their only hope is a journey into the unknown.

Any adolescent lion that climbs the Crater walls quickly discovers that the outside world is much less promising than the Crater itself. Wildlife is scarce in the highlands. The most plentiful prey are Masai cattle, and lions that develop a taste for beef soon wind up on the end of a spear.

Strange lions occasionally show up on the Crater floor, but they don't last long. No immigrant has successfully settled down here in over twenty-five years. The Crater is a lion factory bristling with so many teeth and claws that the traffic is all one way: up and out.

After finishing our morning circuit, we stop off at the Lerai forest to collect firewood. Bright yellow bark glows against the dark backdrop of the Crater walls; the impenetrable stand of fever trees is confined to a large, triangular swamp. Many of the young trees have been pruned by hungry bull elephants that come here from Lake Manyara. Female elephants never come down to the Crater floor, presumably because of all the lions and hyenas. Bull elephants are too big to be bothered by anything except each other, but they aggregate peacefully here, browsing constantly, and recharging their batteries before returning to the wars.

We drive to camp with a stack of rotten branches piled high on our roof rack. Rob and Erin arrived at dusk last night, and we pitched our tents two hundred yards downhill from the Lerai Cabin. Rob is in good spirits after his trip to Australia, but he is not very confident about getting the job.

Sitting around the campsite after lunch, we enjoy the birds hopping around on nearby bushes and tree trunks; redheaded woodpeckers, yellow weaverbirds, shrikes, fire finches, guinea fowl, and metallic-green cuckoos. I am essentially back to normal today, but I appreciate the chance to sit quietly and recuperate. Then Rob delivers the mail from Nairobi. A couple of faxes, one dated 6 November from Anne. It snowed thirty inches in Minneapolis at Halloween. The second fax, dated 5 November, is from an assistant editor at *National Geographic* who has suggested a number of changes to my article on the Crater lions. Hmmm.

The *Geographic* article tells a story that began thirty years ago, when Henry Fosbrooke was still conservator and the Ngorongoro Conservation Authority was very young. Little was known about the Crater, or its lions, when the heavens broke open one dark November evening to unleash a plague of biblical dimensions, a plague whose consequences persist to this very day.

In northern Tanzania, the short rains of November and December are usually followed by a two-month dry spell; then the long rains begin in March. But the rains that began late in 1961 were unusually heavy and continued without a break until the end of May. The soil of the Crater floor became the perfect breeding ground for *Stomoxys calcitrans*, a biting fly that looks like an ordinary housefly but bites like a tsetse. Breeding at their theoretical maximum for seven months, the flies filled the Crater with vast, bloodsucking swarms.

A dense cloud of *Stomoxys* flies can bleed an animal to death within a few days, and the Crater flies did kill a few Masai cows and an old bull eland. Mostly, though, the flies tormented the lions. The stricken lions became emaciated and covered with festering sores. Some tried to escape up trees; others hid in hyena burrows. Within weeks, the Crater lion population had crashed from about seventy animals to less than ten.

The *Stomoxys* plague ceased once the rains finally abated in June of 1962, and Henry Fosbrooke wrote a short report about the lion population crash for the *East African Wildlife Journal.* During the next four years, Henry photographed the surviving lions and kept notes on the population's recovery. Twenty years later, I knocked on his front door at Lake Duluti. I had come to ask if he could help reconstruct the precise history of the Crater lions from the *Stomoxys* plague to the present day.

When Anne and I first came to Ngorongoro in 1979, we looked upon the Crater lions as merely the spoiled children of opulence. We had inherited the lion studies from David Bygott and Jeannette Hanby in early 1978 and had spent our first months learning our way around the Serengeti before finally coming to the Crater. After the Serengeti lions, which we had watched coping with the annual hardships imposed by the wildebeest migration, the Crater lions seemed like fat cats on Easy Street, their next meal always in view. Our only sympathies lay with the subadults, animals excluded by the very system that had guaranteed their constant comfort during childhood.

But by the early '80s we realized that the Crater lion factory was powered by only a few individuals. Although the Crater holds a large number of lions per square mile, most youngsters are forced to disperse, and no newcomers manage to settle here permanently. The breeding population is therefore very small, and very exclusive. And if it had been this exclusive from the time of the biting fly plague, the population might be considerably inbred.

As more and more of the wild places throughout the world are carved into Gombe-sized chunks, many animals are left isolated in small populations. Close inbreeding becomes inevitable, and high levels of inbreeding reduce fertility and genetic variability. A population lacking in genetic variability is less able to survive environmental change over time and, in the short term, may be more susceptible to infectious disease. If each animal carries precisely the same vulnerabilities to disease, one successful bug can devastate an entire population.

Like Gombe, most small wildlife reserves have only recently been surrounded and cut off by the burgeoning human population. But Ngorongoro is naturally isolated, and by investigating the genetics and reproduction of the Crater lions, we hoped to see what the future might hold for other small, isolated populations. We therefore contacted Steve O'Brien at the National Cancer Institute and his colleagues at the National Zoo in Washington, DC. O'Brien had recently discovered that cheetahs had undergone a prolonged period of intense inbreeding within the past few thousand years. Every cheetah in the world now shows evidence of this history, including high levels of sperm abnormality among males and extremely low levels of genetic variability among both sexes.

In 1985, we arranged a visit to Tanzania for Steve and his collaborators—David Wildt, a reproductive physiologist, and Mitchell Bush, a veterinarian and anesthesiologist. After a week spent collecting blood and sperm samples from the Serengeti lions, we arrived in Ngorongoro one morning to find six male lions spread out along the lakeshore in the center of the Crater floor. Mitch used a dart gun to anesthetize the males in rapid succession, and we prepared to collect the semen samples.

An electroejaculator is a thick brass rod connected to an electrical power source. Mitch inserted the device into the animal's rectum, and Dave turned on a portable generator. Even though he was heavily anesthetized, the male would stiffen his back legs, extend his claws, and make a peculiar faint, whining, groaning noise: errr-errr errr-errr errr-errr. Mitch collected the semen in a Dixie cup, and Dave immediately examined each sample under a microscope.

It would be months before O'Brien's lab could perform genetic assays on the lions' blood samples, but Dave was already excited. He could see that the sperm showed a large number of abnormalities compared to samples from the Serengeti lions. Many of the Crater sperm were bent and swam uselessly in tight circles; others had two heads and looked like Mickey Mouse.

So here we were standing next to all these lions lined up on the lakeshore like sausages; one man was staring enthusiastically through his microscope, another was squatting with his Dixie cup while a lion was going errr-errr errr-errr errr-errr. A convoy of ten tourist cars came rolling up, and Steve, who had been spending most of the afternoon smoking cigars and admiring the view, suddenly decided to pull out his video camera and film whatever I might say to mollify these poor tourists who had paid thousands of dollars to come all the way to Africa to see virgin wilderness, and who instead had found some bizarre form of ritualized sodomy being performed on the king of the African plains.

Dozens of horrified faces were poking up out of the Land Rovers: Americans in baseball caps and sunglasses, Europeans with children, Asian women behind their veils. But as I started to explain the rationale of the exercise, one of them interrupted. “Oh, yeah, like in the cheetahs, I read about that in *Scientific American.* This is really interesting!”

So we all survived somehow. The tourists were satisfied, the samples were collected, and the lions awoke from their slumbers eventually to become fathers in spite of their stolen seed. We discovered that the Crater males had consistently higher levels of abnormal sperm than the Serengeti lions, and Steve O’Brien eventually confirmed that they had lower levels of genetic variability as well.

But what was the precise level of inbreeding that had produced these effects? Was the current population descended entirely from the few survivors of the *Stomoxys* plague? Or had large numbers of outsiders settled into the Crater after the population’s decline? I went to visit Henry Fosbrooke in 1986 to see if he could help solve these puzzles.

Henry’s Masai wife served us tea as we looked at his photographs from the early ’60s. Henry put on his reading glasses and spread his old black-and-white contact sheets across the dining table. He quietly told me about the single group of four females that had largely repopulated the Crater. They had lived along the Munge River and had raised large sets of cubs in 1963 and 1965. I listened to everything he said, but my eyes remained glued to his photographs. They clearly showed the whisker spots and ear notches of all four females. Several shots showed two or three females in the same frame, others showed mothers with their cubs. Taken together, they provided a precise record of the foursome’s family composition.

Realizing how much information could be learned from a single batch of old photographs, I was intrigued by the prospect of becoming a detective. If we could somehow assemble enough clues, we might be able to construct a complete family tree for the Crater lions. It seemed like a long shot. Although only a handful of lions had survived the plague, the lion population had reached one hundred by the time David Bygott and

Jeannette Hanby started their Crater study in 1975, and we would have to fill in a ten-year gap. But the Crater is one of the most heavily photographed spots on earth, and the lions are one of the main tourist attractions. Every Crater lion is photographed dozens of times each year; all the data must surely exist, albeit scattered all over the world.

I said good-bye to Henry and his family, drove straight back to SRI, and wrote an article with Anne for an East African wildlife magazine. We gave a brief account of the *Stomoxys* plague and solicited photographs from anyone who had visited the Crater in the '60s.

The response was remarkable. Over the next year, photos poured in from Europe, North America, Australia, and East Africa. We heard from conservationists and businessmen, from retirees and widows, from people who wrote fondly of their holiday in Tanzania twenty years ago.

Meanwhile, we pestered every wildlife photographer we knew, and located an ex-Crater lionologist in the Yukon. An old friend from SRI showed up in Minnesota with a box full of lion slides that he had taken in the mid-'60s. We received hundreds of negatives and prints, sketches, and identification files.

After three years, we had finally assembled enough material to start work. We sat around a large lab table on a hot summer day in Minneapolis and stalked our quarry with a magnifying glass and a slide projector. We stared at the whisker spots of long-dead lions projected on the walls and squinted at sketches and negatives spread out on every available surface.

Within two days we had filled half the span between Fosbrooke and the Bygotts. Fosbrooke's four females had indeed been the ancestors of several contemporary prides, but we still had a long way to go. The biggest gap was between 1967 and 1969. We made more inquiries, issued new appeals. New leads eventually appeared from Kenya and Quebec; a steady trickle of information slowly filled the void.

I took each new batch of film to be printed at the local shopping mall. After picking up the prints, I would tear open the package and stride through the air-conditioned courtyards, staring excitedly at whisker spots while surrounded by jaded American shoppers. A tidy, intricate family tree was emerging. The pieces of the puzzle were falling into place. I felt a constant exhilaration—the gamble was starting to pay off.

Then one day the puzzle was finally completed. We were able to trace the ancestry of every Crater lion back five generations to just seven females and eight males. The seven females were two sets of sisters who had survived the plague, while most of the males were immigrants who entered the Crater shortly thereafter. Only one male had survived the plague, and he was unable to block migrants from the outside world.

The complete family tree of the Crater lions is as complex and incestuous as that of European royalty. Fosbrooke's founding female foursome had raised three successive cohorts of young. Their daughters and their daughters' daughters had established five of the six contemporary Crater prides. After a brief period when immigrant males invaded the Crater floor, the foursome's first two sets of sons formed large coalitions that eventually resided in every Crater pride—most of which contained their sisters and cousins—and prevented any further male immigration from outside. Six of the founders' grandsons dominated the Crater for nearly a decade, fathering almost all of the cubs born on the Crater floor, and again their mates were their sisters and close cousins. Although lions usually avoid mating with close kin, these animals had had no choice—the entire population was just one big family.

We fed the family tree into a computer program written by Henry Rowley, an undergraduate at Minnesota. The program allowed each lion to "breed" according to the marital history of the population and then compared the genetic composition of their simulated descendants with the results of our actual blood tests. Henry's simulations suggest that the real Crater lions have lower levels of genetic diversity than would be expected from a single population bottleneck. We therefore suspect that earlier generations of Crater lions were also victims of the Crater's splendid isolation. Perhaps their earlier loss in genetic diversity had resulted in an extraordinary susceptibility to the *Stomoxys* plague of the '60s. Perhaps the plague actually benefited the population by temporarily opening the door to new genetic diversity from immigrating males.

The overall reproductive rate of the Crater lions has been steadily declining for the past twenty years. All these well-fed females are producing fewer and fewer surviving cubs with each passing year. While this trend may be a result of the cumulative effects of chronic inbreeding, it may also be the effect of high population density. Most organisms breed more slowly as they become more crowded, and the Crater lion population has increased steadily since the plague. Determining the relative importance of each factor is essential to assessing the long-term future of this population. If the lion population spirals downward due to inbreeding, then it will be necessary to import new blood artificially. If the population is merely regulating its own density, then its future may be secure, barring any further ecological catastrophes.

After reviewing the recent lion news with Rob this morning, I am struck by the low number of cubs in the Crater today. The population is no more crowded now than it was ten to fifteen years ago, but cubs are much more scarce. However, even if the cub shortage is due to inbreeding, these lions still produce more young each year than could possibly settle here. Inbred populations do not always go into a fatal decline; the cheetah, for example, persists throughout much of Africa in spite of its genetic legacy.

Introducing new blood into the Crater lion population would be extremely difficult. If we brought in new males, they might panic in an unfamiliar area and leave immediately. Even if they stayed, they might wreak so much havoc killing cubs and chasing out subadults that the population would be further damaged. Any attempt at artificial insemination would require sequestering each female for several days at a time and preventing her from mating with any of the local males. Even then, we would only produce a single litter that might be lost at the next male takeover.

Perhaps the best hope is that immigrant males will solve the problem for us. The *Stomoxys* plague eliminated all but one male on the Crater floor; new males were suddenly able to enter, and they probably came in from the Serengeti. If the resident defense forces are weakened again, new blood might invade once more. Fortunately for the local lions, the Ngorongoro Crater is blessed with a fantastic carrying capacity, but it is so close to the Serengeti that it can only remain isolated during periods when Crater males are strong. Fortress Ngorongoro, though, will always be subject to decay from within.

I wanted to write a general account of this story for *National Geographic* not only because of its implications for the future of small populations but also because it highlighted the extent to which science depends on large-scale cooperation and the good will of the general public. The construction, application, and interpretation of the Crater genealogy was possible only because of the efforts of a dozen experts and the help of nearly a hundred "average citizens" who sent us their cherished photographs.

We have become so used to hearing that our normal lives are destroying the planet—our cars are creating a greenhouse, our air conditioners are depleting the ozone. But here was a case in which average people had unwittingly acted as research assistants. They had documented critical events that could be brought together in a meaningful way. Here was something they might feel good about.

SUNDAY, 1 DECEMBER

After lunch, everyone sits around eating fruit, reading books, writing letters. The campfire is smoldering, and someone has put a kettle on for tea. An occasional gust of wind kicks up the ash and whips the tents. The Crater lake has almost dried up and the shallow water mirrors the afternoon sky. Dust devils glide slowly across the alkaline flats.

Sarah, Pam, and Christine have settled in to their new lives. No longer just visitors, they have become part of the team. With their growing confidence, they will express themselves on any issue, and they are not afraid to yell at me if I do something stupid. Thrilled to be here, their enthusiasm is infectious and focuses my attention on the sheer beauty of the animals and landscape for the first time in years.

These three women are not easily impressed, however, and can be especially critical when it comes to male lions, having formed strong views on male beauty during their weeks in the Serengeti. They have high standards for overall physique, health, and the size of a male's mane, although they disagree about mane color—Pam and Sarah both like black manes, while Christine prefers blonds. Bring them to the Crater, where most males have coarse features and odd dimensions, and their assessments are quick and direct.

"He's so ugly!"

"Squat!"

"Droopy jowls!"

"Pug face with piggy eyes!"

Stretched out on the open ground, legs and paws splayed out in every direction, even their postures are unacceptably awkward.

"Look at him, he's double-jointed!"

"Lopsided!"

Somewhat discomfited by these vehement outbursts, I find myself trying to defend the lions. "Yes, but he's so big—much bigger than a Serengeti male. And there are four of them; they can protect their cubs against anything."

But no go.

"He's *so* ugly!"

"Must be all that inbreeding."

MONDAY, 2 DECEMBER

Standing at the edge of a swamp near Tokitok Springs, a male lion gnaws on the stinking remains of a dead hippo. The rest of the pride have disappeared into the dense green reeds. Nearby, dozens of jackals and vultures are awaiting their turn at the carcass. The lion looks as if he is about to burst—all that remains of the hippo is bone and hide. The sun is high, and he wants to join his companions in the cool of the reeds, but he isn't ready to relinquish the tattered carcass.

He walks two steps away from the scraps, but an impatient jackal sneaks in at the crucial moment. The lion, incensed, chases the little pest away and decides to take the mess with him. Despite being completely gorged, the male plants his feet, lifts one end of the six-hundred-pound pile of skin, bone, and crud, and drags it toward the swamp, an almost mythical embodiment of greed.

After lunch, we put on our hiking shoes and start to follow a wide cattle trail from the Lerai Cabin up to the Crater lodge. The Masai water their stock on the Crater floor during the dry season and drive their herds from one highland pasture to the next throughout the rains. The trail braids through the blue-green scrub of the upper Crater wall, recording the reluctant separations of countless herds of cattle.

We encounter no hazards on the way up, but the steep and dusty path has been scored by the footprints of a very large bull elephant. His tracks are not only enormous but wrinkly, furrowed like a worried brow. Elephants have also gouged out sections of a nearby cliff-face with their tusks, seeking salt in the volcanic soil.

Birdsong along the way reminds me of various snippets of music. The Debussy bird with its theme from the second Arabesque, now the shrike with its ridiculous call from the "Ride of the Valkyries," then the bird with a single bar from Sibelius—yes, this is definitely Sibelius country: soaring, prevailing, triumphant against all adversity.

Sipping our cold drinks at the lodge, we find ourselves separated from the outside world by the large picture windows in the dining room. After a few days in the brown, dusty fishbowl, we are all relieved by the sight of the hazy blue volcanoes to the north and the fog-drenched forest along the eastern rim. But from the inside, they are just flat pictures on the glass.

On our way outside, the receptionist hands us two handwritten messages dated 28 November. The first is addressed to me from Pam and Sarah—they have given up waiting for Christine and me and have gone down into the Crater. The second is for Pam and Sarah from Christine—she has been stranded somewhere illegible. Could they please come and fetch her?

TUESDAY, 3 DECEMBER

We sit perched around the campfire in the grey light of morning. The sky has brightened, but the sun won't be rising above the Crater rim for another half hour. The dawn air is still cold, and a chilling breeze tumbles down the Crater wall. Our kettle starts to boil, and the caffeine addicts stand poised, anxious to wake up.

Driving back toward the hippo kill, we encounter two bull elephants walking along the road. Slowly flapping their ears, the elephants head for the shade of Lerai forest after a night of foraging in the swamps. The soles of their feet balloon outwards with each careful step. We leave the track, yielding the right-of-way to their great lumbering bulk. To our left, a rhino mother and calf stand grazing in the open plain. The rhinos are survivors of another plague. All but ten were slaughtered by poachers during the '70s and '80s. Dozens of crumbling rhino skulls are stacked beside the Lerai Cabin, each one gashed where their horns had been hacked off.

At Tokitok, the hyenas have disposed of the evidence, and the only trace of the hippo carcass is a stain in the grass. About a hundred yards away, three male lions lie in the open, surveying their domain. Their territorial boundaries are all visible, and no trespassers are in view. They soon collapse into a peaceful sleep, oblivious to the noisy insults of three very particular women sitting in a red Land Rover.

We skirt the north shore of the lake until we reach the mouth of the Munge River. The Munge follows a winding, tree-lined course before entering a large swamp, then reemerges as a narrow, reed-lined channel before fanning out on the lakeshore. Wildebeest and zebra come to drink from the fresh open delta before it mingles with the alkaline waters of the lake. The lake has almost disappeared in the past few days; stream water quickly evaporates on the snow-white pan.

Driving along the banks of the Munge, we spot two female lions walking furtively in the tall grass. They are thin and, aha, they are LKT and LKU. They continue toward the swamp, seeking cover and avoiding any possible confrontations with another pride.

LKT and LKU were both born in the Lake pride, but their story provides a unique insight into the complex nature of female sociality. LKT was born a few months before LKU, and she was raised in a large crèche formed by five females. LKU's mother only associated with this crèche on a part-time basis—her small cubs were unable to feed with so many older cubs scrambling over the same carcass. When a seventh female gave birth a few months later, LKU's mother withdrew her from the overcrowded classroom and formed a second crèche. The two crèches foraged apart, splitting the pride in two. When the main pride was taken over by a new set of males, LKT was evicted along with her cousin, LKR.

Meanwhile, LKU and her cousin, LKW, were orphaned, and, at first, the two pairs tried to establish separate territories. LKU and LKW attempted to settle near the Lerai Cabin, and LKR and LKT moved toward Seneto Springs. But both pairs were rebuffed by larger prides and eventually returned to their original range to fight for control of the Munge River delta.

Having battled each other for years, their conflict ended in a stalemate. Neither pair could defend the delta against the much larger Lake pride.

When LKW disappeared two years ago, LKU finally declared a truce and teamed up with LKR and LKT, six years after they had separated. LKR died last year, leaving LKT and LKU as the sole survivors of their respective crèches.

Pride divisions are almost always permanent; once female lions are separated for more than a few months, allies become rivals for the rest of their lives. LKU and LKT are the only two females who have reunited after years of hostility.

Prides seldom flourish unless they contain at least three adult females. But in contrast to male lions, solitary females almost never team up to form prides with unrelated companions. The only way most females enlarge their pride is by recruiting one of their daughters. Why don't females follow the males' example and team up with any other solitary they meet? Female lions would always benefit from joining forces. There is no queen enforcing sterility on her subordinate kin, like termites or naked mole-rats. Why not escape the horrors of solitary life by finding a companion? Why associate only with family or with no one at all?

Females have very few opportunities to meet suitable companions. They are reluctant to leave their mother's range, even after departing from their mother's pride. Thus, if a solitary will stay only in a familiar, and familial, area, her only candidates for companionship will be her relatives.

Solitary females do sometimes form brief friendships with strangers during forays to fields of plenty. But when the prey moves on, the females return to the separate loneliness of their childhood homes. In all of Africa, there have been only two documented cases of solitary females joining nonrelatives to form permanent prides. Both cases involved females that had been forced to abandon their original ranges by extreme drought or human disruption. Only when forced to settle away from her family home will a solitary remain forever in the company of strangers.

Why should females be so reluctant to move away from home? As principal caretaker of her cubs, a female lion needs to know the safest dens and the best locations for finding food during hard times. A young female giving birth for the first time knows only one good den—the den where she was born.

Furthermore, the world is hostile and crowded. No one welcomes a young lion looking for a place to call home. However, if the castaway youngsters are your own offspring, you should be more tolerant of them than of strangers. These are your children, after all, your stake in the next generation. Allowing the kids to camp in the backyard may not guarantee them a successful start in life, but it provides them with a safe refuge until a neighbor's house goes on the market.

LKT and LKU are the exception that proves the rule. The only two females that have reunited in the Crater also shared the same family homestead and desired the same small stretch of river. United by a struggle against common enemies more powerful than themselves, they finally overcame their own animosities. They have pulled together on the battlefield.

They are ten years old now and have never reared cubs, but they have managed to stay on the Crater floor. They are shoved around constantly by the larger prides, but at least they have companionship. And who knows? Maybe they will be the lucky ones that survive the next plague.

The night is cold and windy. We sit in a tight circle around the blazing campfire, our eyes held captive by the flames. Erin steers the general conversation from one topic to the next: pop stars with AIDS, the British versus the Australians, the condition of the Aborigines. She has relaxed considerably since our confrontation in the Serengeti, but we are all still careful to watch our step.

I sit to one side, listening quietly as an hour passes and the flames slowly flicker and die. Then the discussion turns to the latest news from the World Service of the BBC. The Soviet Union is collapsing. Ethnic violence is spreading throughout Eastern Europe. Nationalist movements are proliferating around the world.

In the context of the news, the stories of coalescing females suddenly take on the resonance of a parable.

Leave two females with their mothers, and their first allegiances are to their own families. Separate them from home and place them on neutral territory, and they will assimilate as convincingly as any two immigrants to the New World.

According to the news, the strategic alliances of East and West have crumbled with the lost threat of a hostile ideology. Being as unwilling to reforge our political boundaries as any other species, it's just too bad that we have run out of Minnesotas, as Dr. Rosling would have observed.

WEDNESDAY, 4 DECEMBER

Lunch is over, the cars are loaded, and we are ready to bounce back to Seronera. Christine releases the brake, and the Suzuki starts rolling downhill. She releases the clutch, and the engine fires immediately. Pam and Sarah hop into the Land Rover, put on their sunglasses, and prepare to follow at a safe distance.

At the base of the ascent road, we put the cars into four-wheel drive and brace for the climb. The lightly packed Suzuki scoots along the rocky, twisting road with ease. Sweeping vistas invite the incautious eye to linger on one last view of paradise. Take more than a brief glance, however, and risk a fast flight to nowhere.

Surrounded by dry yellow plains, the Lerai forest resembles a green bikini bottom hastily discarded on a beach. The lake has dried out completely, and the Munge glides into the alkaline flats with the grace of a cobra. All around us, the Crater walls testify to the improbable order of planetary forces. Our straining vehicles sputter up the slopes like two fleas on the flank of Leviathan.

Winding, climbing, now we're dusty, now slick with dripping dew. The engine grinds, the bodywork creaks, and the temperature gauge is off the scale. Then release, as we reach the cool air of the summit and the intersection with the main road. The tourist minibuses have started their own migrations for the day, some heading toward Seronera, the rest for Manyara. Gaudily dressed Masai pose eagerly by the roadside. One shouts in English, "Give me two hundred!" while her companions bounce their beaded necklaces, hopping up and down.

The western sky is clear and blue, and we soon see the Serengeti, partially hidden by the haze below. The barren peaks of the Gol mountains burst through the short grass plains like bare fists crashing through a taut rubber sheet. Rips in the fabric turn out to be dry, shallow riverbeds lined with flat-topped acacias.

A Masai boy quietly tends his herd of goats on the outer slopes of the Crater. Wearing only a tattered cloak, he has an unmistakable dignity. He has no interest in us or the modern world; he would be offended if someone were to take his picture.

Many years ago, a *National Geographic* film team went to the Serengeti to make a documentary about the Serengeti Research Institute. However, the SRI scientists refused to cooperate, and the film team was left high and dry. While kicking their heels in Seronera, they met a young Masai tour guide who had been born in one of these villages in the Crater highlands. An exceptionally bright young man, Ole Saitoti had been chosen by his father to be the one member of his family who would receive an education, the one son to lead his people into the future. The filmmakers centered their documentary around him and brought him to America, where he eventually received a college degree.

We met Ole Saitoti shortly after he came home from America. We gave him a lift from the Crater rim to the Serengeti. He was charming and enthusiastic, but he had become attached to the comforts of civilization.

"I went to stay with my parents when I first got back. But their hut was so dirty! And the smell! Living with cattle and no baths for months on end!

"You see these people walking around these highlands in their sandals and their red blankets. I'll tell you, those blankets are too thin. It's cold up here! There's frost at night! And the food. I like my *ng'ombe* (beef) and *mbuzi* (goat), but I want vegetables, too. And bread!

"It's rough, but I'll tell you one thing, I'm a warrior now—and a warrior can have any woman he wants."

About ten years later, I was in Minneapolis watching a newly released television documentary. Halfway through the film, Ole Saitoti suddenly appeared, preparing for an important tribal ceremony. He sat uncomfortably while his head and eyebrows were being shaved.

The narrator solemnly described how Saitoti must pass on the privileges of the warrior to the next generation and prepare to become an elder, an *mzee.* Bald and shivering beneath his flimsy blanket, he was led out to a small clearing and instructed to sit on the viscera of a freshly slaughtered cow with his wife and aging mother.

I don't think I have ever seen anyone look more miserable.

We arrive back at SRI just before dark. In the backyard of the Lion House I light a fire under a fifty-gallon drum connected to the plumbing system and fill a second drum with enough water for four hot baths. The flames roar up into the night, and the lower barrel starts to creak and groan. In the bathroom, the water barely trickles from the showerhead, but nothing can compare with the sheer joy of being clean in the Serengeti.

I walk over to the rosemary and the pure white kitchen of Pam and Sarah's house. The smell of supper permeates the air, the quiet music defeats all sense of exhaustion. Pam and Sarah, each in their own way, tell me how much driving across the plains felt like coming home.

The plains. The Serengeti plains. Plains spawned by great volcanoes. Home to this giant clockwork of animals chasing an endless cycle of life and death.

Home.

SERENGETI / THURSDAY, 5 DECEMBER

Peter Arcese and I are sitting on the verandah. Sunday, experts on every major component of the Serengeti ecosystem will gather and spend four days assembling a vast computer program. We hope to learn something about the long-term prospects for the Serengeti by integrating the many interactions between climate, soil, vegetation, animals, and human activities in and around the park. Several dozen scientists will be arriving at SRI in the next few days, and Peter, as co-organizer of the symposium, has to worry about where to house them and how to power all their portable computers.

Peter studies antelopes in the far north of the park, near the Kenyan border, where rainfall is heaviest. The northern Serengeti is a mosaic of streams, scrub, and grassland, a critical refuge for the migratory herds during the dry season and home to large numbers of resident species year-round. However, the area's richness attracts a particularly ruthless class of poacher. Game posts near the northern boundary have been attacked twice in the past year, and rangers have been murdered. Peter had to move down to SRI last year after his own research camp was raided.

The predominant tribe in the north is the Wakuria, and they are renowned for their ferocity. The Wakuria lost access to their traditional hunting areas when the Serengeti was enlarged in the early '60s. They declared independence in the mid-'70s, pulled down the Tanzanian flag, and replaced it with a leopard-skin banner. Their insurrection was soon crushed, but the Wakuria retain a bitter resentment toward the park and its managers. They express their anger through a general lawlessness and disregard for all boundaries.

In addition to poaching for meat and trophies within the park, bands of Wakuria regularly cross into Kenya, raiding and robbing wealthy tourists in the Masai Mara. They have also started attacking travelers in the northern Serengeti.

A party of German tourists was recently assaulted just south of the Kenyan border. They were forced to march several miles from their car, then they were robbed, stripped, and abandoned. On their walk back to the road, the tourists found a slip of paper that had been dropped by the bandits. It was a permit allowing a certain man to herd his cattle across the park. Barefoot and angry, the Germans flagged down a car and took the cattle permit to the police in Seronera. The police arrested the permit holder and recovered the tourists' belongings.

Shortly thereafter, the Tanzanian government dispatched a paramilitary unit to clean up the whole region. Undercover agents had already infiltrated the villages, so the troops knew the names of the major poachers. The soldiers summoned the worst offenders to the village squares and told them to hand over their weapons. If a suspect refused, one of his ears would be lopped off. If he refused again, he would lose the other ear. Seven hundred rifles were eventually confiscated. Most of the villagers welcomed the soldiers—the area had become so lawless that everyone lived in fear.

Christine has finally examined the leaking baboon samples that have been stewing in my canvas bag since we left Dar es Salaam. Each vial was labeled with the name of the baboon and the date on which the specimen was collected, but many of the paper labels were shredded during the bouncing journey back to Seronera. Christine possesses a considerable sense of hygiene and order, and the unspeakable mess has been more than she could face. It takes most of the afternoon to restore calm and decipher the remnants of the three hundred labels. In the end, only ten have been rendered completely useless.

Pam and Sarah collected a dozen more lion samples while we were away, but no one fell for the last few ox hearts, and according to Rob there is no male fern extract in Nairobi. Although Christine's work would have been simplified if she could have obtained adult worms, she can get by with the parasite eggs in ordinary feces. Pam and Sarah will be sending lion samples to her for the next two years, and we are optimistic that Erin will eventually hand over her forty hostages.

FRIDAY, 6 DECEMBER

The woodlands around Seronera have changed dramatically over the past twenty years. Age and elephants have decimated the stately umbrella acacias that once defined the landscape, and thick stands of brushy saplings have sprouted in their place. Kept down for decades by wildfire, elephants, and giraffe, the rising generation of scrub has at last been liberated by the collapse of the elephant population and a policy of controlled burning. The scrub is ten to fifteen feet high, finally tall enough to withstand flames from below and escape the tongues of the tallest giraffe. Someday the region will be beautiful again, but at the moment it is an unsightly tangle that scratches the sides of our car, mangles our antenna, and conceals even the largest pride of lions.

Sarah is driving; Pam is listening on the headphones. Beep beep Beep Beep BEEP. We are closing in on the Sangere/Northwest pride. "There they are! All of the females are here and, look, there's History. He's with another male today—is it Progress?"

But I'm doubtful. "Drive past those bushes and take a closer look."

"See, he's got that scar in his nose, and the notch in his right ear."

"The scar does look similar, but History had two notches in that ear. This male has only one notch, and the spots on the left side are definitely wrong."

"We couldn't get a good look at his left side."

"History died two years ago. Jon and Karen found his collar. That second male looks familiar. Yes, here's his card—Golfball. He's got that little lump on his side, and see the scar on his lip and the tatty ear? That means that your friend History must be Buffalo Bob."

"Oops."

"How embarrassing."

"That's okay. You've been such fast learners, you've become a little too confident. Mix-ups happen, and this was a good way to learn. Anne and I made the same mistake when we first started. A tailless male showed up just a few weeks after we arrived. His tail had been bitten off about four inches from the base. We were impressed by how quickly he could travel from place to place, until we finally saw him in two places at once. Two tailless males had shown up out of the blue in the same week. Later, we were amazed we had ever mixed them up—they had different-colored manes, and their spots were totally different."

"It's embarrassing just the same. Especially to resurrect someone who has been dead for two years."

"Well, yes, it is always a good idea to be skeptical when you think you've seen a ghost."

With pencil and paper on Pam and Sarah's verandah, I prepare my talk for the symposium. Rain fell here last week; the hills and trees are trying to turn green. The sky is crystalline; sunlight slices the air like a knife. Then that voice again, "*Hodi.*"

It is Kisiri, just back from court. The judge has delayed his decision until January, and Kisiri is truly crestfallen. His case has been dragging on for months. He has been ill with amebic dysentery and malaria. His malaria is not responding to medication, so Sarah and Christine run inside to consult a medical guidebook called *Where There Is No Doctor*. They give him the newest drug from Europe; he should recover in a few days.

But Kisiri's spirit is broken. Not by his poor health but by the loss of his good name. If his honor is taken from him, his life will be done.

I can only repeat my assurances that if he is acquitted, I will employ him and pay off his debts. The case against him is not strong, but I cannot pass judgment. If he is judged guilty, we cannot reward him. "I will be going back to America in a few days, but please, please come see my students if you are acquitted."

Half an hour later, Teddy arrives for a visit. She is the sister of our former nanny and came to SRI a few years ago to work as a housemaid and cook. Last year, she became embroiled in a nasty furor over the crime and corruption of the SRI staff camp. She had reported several incidents to the police at Seronera, and the rest of the SRI staff demanded that she be booted out of the park.

She resisted for months, and her case was taken up by the scientists at SRI, but to no avail. She was ordered to leave last month and has only returned to ask for bus fare to Kigoma. She will be working for Alan Root's team, who are filming fish in Lake Tanganyika.

While she is here, Teddy wants to make sure she is replaced by a woman she trusts, and she introduces her friend, Shida.

Shida is Swahili for trouble.

SATURDAY, 7 DECEMBER

The troops have assembled, but there is grumbling in the barracks about the objectives of the campaign. The symposium starts tomorrow, and we are expected to produce a complete computer model of the Serengeti ecosystem in just four days. That goal would be hard enough to achieve at the best of times, but the scientists here are a diffuse group, a collection of rebellious free spirits suspicious of centralized planning, experts reluctant to address questions outside their narrow specialties.

Tony Sinclair is the real *mzee* of the Serengeti research community and the driving force behind the workshop. One of the last of the original SRI scientists, he was born in Tanganyika (his father was in the colonial service), raised in New Zealand, and educated at Oxford. He teaches at the University of British Columbia and comes out to Tanzania for a month or two each year. His research is still inspired by SRI's original mandate.

The Serengeti Research Institute was established in the mid-'60s to address two central questions. How does the Serengeti support such large numbers of animals? How can so many different species coexist?

Funds poured in from every direction, houses and laboratories were built, a director was hired. Senior scientists were gathered from around the world, and each was assigned his own role in answering those two central questions. George Schaller was asked to study the lions, and Sinclair came to work on Cape buffalo. There were geologists, ecologists, botanists, parasitologists, and veterinarians; Americans, Australians, Brits, Germans, Dutch, and Canadians.

SRI ran like a dream. An airplane could fly you to any corner of the park at a moment's notice, fleets of Land Rovers swarmed across the woodlands and plains. A lorry went to Arusha every week for supplies, there were daily scheduled flights from Nairobi. Birthday party coming up? No problem—have some ice cream delivered from Kenya. Lovely to see you, and this is my mother, *Lady* Trumpington-ffines. Three-course dinners, cutlery adorned with the family crest. Servants, swimming pools. The place was jumping: drunken picnics, wife swapping, gliding, tennis on Sundays. It was the '60s, all over the world.

Not everyone became immersed in these distractions, however, and comprehensive studies were carried out on carnivores, ungulates, birds, grasses, trees, fire, rain, soils, and disease. The place was inexhaustible, and prospects for research seemed endless. It was the last place on earth where nature ran free, and there was no better place to be.

George Schaller's lion study turned out to be a true landmark, not only because he worked out so many aspects of their social lives, but because his findings made it clear that lions, leopards, cheetahs, and wild dogs did not limit the population sizes of their prey species. It was not necessary to persecute carnivores in order to ensure large populations of herbivores; the Serengeti could persist forever as an unspoiled spectacle of predators and their prey. Although we might take this notion for granted now, park wardens used to shoot every wild dog they saw and even eradicated spotted hyenas from several reserves.

George has set out to learn as much as possible about endangered species before they disappear from the wild. Before coming to the Serengeti in 1966, he had already studied gorillas and tigers, and after the lions, moved on to jaguars, snow leopards, and pandas. Just before George left the Serengeti in 1969, an English scientist named Brian Bertram asked to continue the lion project. Schaller was initially reluctant, because so many of the park's inhabitants were still unstudied. But Brian was determined to study lions, and Schaller relented. Bertram would monitor the prides around Seronera but base himself in a different part of the park. By undertaking parallel studies elsewhere, Brian could confirm whether George's findings were representative of the entire Serengeti.

Brian worked north of SRI, but his most important findings resulted from extending Schaller's study at Seronera. Brian was the first to realize that cubs always vanished shortly after the arrival of new males in a pride. He and Schaller had both seen incoming males kill cubs, but the habitual loss of cubs implied that infanticide was a deliberate behavior rather than an aberration.

Other trends would undoubtedly emerge within the next few years. So when Brian left in 1973 he ensured continuity by handing the study over to Jeannette Hanby and David Bygott, the Hanbygotts. David and Jeannette arrived in 1974 intending to continue the Schaller/Bertram arrangement and to conduct most of their research in a third part of the park. But when they went to the western corridor, they found that the lions had been decimated by poaching and were much too shy to study in detail. They decided instead to contrast the Serengeti lions with the perennially well-fed animals in Ngorongoro Crater.

In late 1977, Anne and I were sitting in the attic of the Hakusan Nature Center in Japan. After writing up our Gombe research, we had received a two-year grant to study Japanese monkeys. We went to Berkeley for an intensive three-month course in Japanese. Our plane from San Francisco to Tokyo caught fire somewhere over the Pacific and made an emergency landing in Honolulu.

Once in Japan, we found ourselves the victims of political unrest in Ethiopia. Large numbers of Japanese primatologists had been forced to leave Ethiopia when a Marxist regime overthrew Emperor Haile Selassie. The Japanese scientists needed somewhere to work, and suddenly all the openings we had expected to find at Japanese research centers were closed.

Our research has always required a detailed knowledge of the animals' personal histories. Are those two companions close kin? Was that male born in this troop or did he enter from elsewhere? How old are they? How many children does she have? Japanese scientists had started to study monkeys in the '50s and had constructed extensive genealogies for a dozen troops. But our hosts were too polite to tell us that we wouldn't be permitted to study these well-known groups.

Instead, they sent us on an extended tour of southern Japan. "The keeper knows all the monkeys. Go there and you will see." Monkey troops are provisioned by rural villages as tourist attractions in Japan. The monkeys spend all day in bare clearings eating wheat or potatoes. Platoons of Japanese tourists file out of their buses to have their pictures taken with *Osaru-sama*, the sacred monkey. Sometimes the local villagers learn to recognize each individual monkey, but mostly they don't.

After visiting a dozen sites we finally faced up to our predicament. We decided to cut our losses and spent a month studying monkeys in a remote mountainous area called Hakusan. Nothing was known about the monkeys' personal histories, but this was the only tame troop in Japan that had not been excessively provisioned. We would try to learn something from a short study and use the time to seek out another species with bona fide family trees.

We stayed in the Hakusan Nature Center. The center had been closed for the winter. Hakusan is one of the snowiest parts of Japan, and there are frequent avalanches. We would only be permitted to stay for another few weeks. The staff had already moved back down to the village in the lower valley. We lived in a small apartment in the attic of the unheated building, where the public toilets were set to flush every two minutes to prevent the pipes from freezing. We sat on tatami mats wearing thick parkas on top of our kimonos, writing letters to friends and colleagues in Europe, America, and Africa. We wrote to the Hanbygotts, asking if there were any openings in the Serengeti. The snow started to fall.

By the time we reached SRI in early 1978, there was no longer any attempt to coordinate research in the Serengeti. SRI had fallen into complete disarray after the institute had been Africanized, and once the border with Kenya closed in 1977 no one wanted to know about Tanzania or the Serengeti. The government had hardened its socialist ideology. Shops were empty, imports were virtually forbidden, research permits were practically unobtainable. We had to have special permits to own a car, drive on Sunday, travel to Kenya. Provisions were scarce, and there were cholera epidemics throughout the country.

In 1979, Tanzania went to war with Uganda. Idi Amin resented Tanzania's sympathetic attitude toward his socialist predecessor, Milton Obote. Amin declared his intention to annex a corridor of Tanzanian territory all the way to the Indian Ocean, then invaded a disputed border zone. The Tanzanians fought back, met surprisingly little resistance, and drove all the way to Kampala. They won the war and replaced Amin with Obote, but devastated their own economy. Tanzania's roads and bridges were shattered by the transport of heavy equipment to the war zone; broken-down vehicles were abandoned in droves. The victorious troops returned home to reap no reward. Disgruntled soldiers stole weapons from their armories, selling them to bandits and poachers. Knife-wielding gangs in Arusha and Dar were suddenly armed with machine guns.

For nearly a year, Anne and I were the only functioning scientists at SRI. Jane Goodall's husband, Derek Bryceson, was Director of National Parks, and he helped us to import a car and obtain research clearance. The final survivors of SRI's golden age taught us how to obtain permits to cross the Kenyan border. We met Barbie Allen in Nairobi, and she kept us supplied with essentials, even during the bleakest times. During the war, we spent as much time as we could out on the vast emptiness of the Serengeti plains, avoiding anyone who might want to confiscate our vehicle to transport troops to Uganda.

Through all this, the African staff at SRI were supposed to monitor the distribution of wildfires, read the many rain gauges scattered around the park, and record the routes of the migration. But their vehicles frequently broke down. Most of the staff had marketable skills, and they eventually found good jobs in Nairobi or Arusha. The few dispirited Tanzanian scientists were allowed only five gallons of fuel each month, enough to drive to Seronera for supplies but not enough to conduct research.

Tony Sinclair came out every other year to survey the wildebeest. After completing his Cape buffalo research, he had started projects in the Yukon and Australia, but he took over more and more duties in the Serengeti, eventually organizing biennial aerial surveys to census the major ungulate species.

Two other Western scientists also played a major role in revitalizing the research institute. In 1974, Sam McNaughton of Syracuse University initiated long-term studies of grazing and grassland productivity in the Serengeti. His support of Tanzanian students during the hard times demonstrated that western scientists could contribute to the development of a Tanzanian research community. In 1978, the Frankfurt Zoological Society sent out Markus Borner, who was based on a small island reserve in Lake Victoria. Frankfurt provided Markus with a plane, and he soon started flying the Serengeti censuses with Tony. Moving to Seronera, Markus eventually resurrected the Serengeti monitoring program.

By the time Tanzania reopened the border with Kenya in 1984, SRI had started to revive. Long-term projects on cheetahs and hyenas had been resumed, as well as a succession of shorter-term studies of the ungulates and the vegetation. With the recent changes in the Tanzanian economy, life is quite comfortable here again, and the place is packed. In addition to all the scientists, SRI houses an education unit and a large-scale regional development project.

But everyone is here for his own reasons. These are all experts without any need of help from you, thank you very much. It will take someone with utter dedication to a larger cause to convince us all to work together, if only for a few days.

SUNDAY, 8 DECEMBER

Nineteenth-century accounts of the Serengeti described a region of almost unbelievable plenty, of low, flat hills black with animals, of herds stretching across vast horizons. But then the rinderpest virus reached Africa, carried to Ethiopia in 1889 by cattle shipped from Italy. Rinderpest is a disease of ruminants that kills up to ninety-five percent of its victims. The virus induces violent diarrhea and vomiting, spreading itself to everyone that dines from the same grassy table.

Rinderpest swept the continent, traveling up the Nile, devastating East and Central Africa, reaching South Africa within seven years. No one knows how many animals died in the Serengeti, but the numbers must have been staggering. Wildebeest, giraffe, warthog, buffalo, and gazelle were all hit hard. Ungrazed grasslands turned into bush, and tsetse flies and sleeping sickness soon followed. People were crushed by famine. Lions sought alternative prey, and many became man-eaters.

Waves of rinderpest rolled across Africa for the next fifty years, but it had become a childhood disease—survivors are immune to further infection, and only the calves are susceptible. By the 1940s the Serengeti was the last major reservoir for the virus in Africa. The Serengeti still held large herds of wildlife, and it was surrounded by cattle. Sufficient numbers of calves were born each year to provide a pool of vulnerable hosts; once a young victim had sprayed the ground with virus, another calf would feed from the poisoned plate.

In the late '50s domestic stock around the Serengeti were vaccinated against the virus. By themselves, the wild herds were too small to sustain the disease, infected patches of ground remained ungrazed. Unable to spread, the virus soon died out.

Liberated from the rinderpest, the Serengeti wildebeest increased from a quarter million in 1961 to one and a half million in 1978; today there are nearly two million. Populations of buffalo, giraffe, warthog, and gazelle also increased severalfold. The low hills swarm once more. One way or another, the herds stretching out across those vast horizons are the reason we have all come here today.

We are all crowded around two long tables in the SRI library. The morning sun shines in through a bank of windows. Shelves of dusty books line the room: textbooks and monographs, checklists and dissertations. Rows of journals have been stacked along the wall: *The African Journal of Ecology, Animal Behaviour, Ecological Monographs, Zeitschrift für Tierpsychologie.* The library is impressive at first glance, but few of its holdings were published after the mid-'70s.

Tony Sinclair is standing beside a hand-painted blackboard propped up on a makeshift easel. A decaying movie screen hangs to one side. A portable generator putters away outside the building, ready to power the slide projector. Tony outlines the goals of the symposium.

"Today we will hear a dozen short talks about the relevance of each person's work to the dynamics of the Serengeti ecosystem. Tomorrow we will start building the computer model, in which we will integrate the knowledge that you have all worked so hard to acquire. You will be splitting up into a series of specialist groups, each with your own computer. A professional program analyst, Ray Hilborn, will be working with us to combine our separate perspectives into a coherent whole."

Most of the audience is enthusiastic, but a few people are openly skeptical, young turks embarrassed by the simplicity of their elders. The essence of the Serengeti is far too complicated to be captured in only a few days. But Tony persists, smiling determinedly. "You'll see."

The day's parade of speakers begins.

Martyn Murray: The Serengeti is famous not only for the abundance of its wildlife but also for its diversity. Over a dozen antelope species somehow manage to coexist, all eating the same grasses. Why don't a few species exclude the rest? In the '70s, several SRI scientists hypothesized that the major migratory species coexisted by specializing on grasses in different stages of growth. Zebra are bulk feeders, preferring mature grasses. Wildebeest seek any green grass. By mowing down the longer grasses, the two larger species create fresh swards of the short green grass preferred by the Thomson's gazelle.

But these three species are as different as horses, cows, and sheep. How do more similar species coexist? Wildebeest, topi, and kongoni are all the same size and share a recent common ancestor, but they too divide up the resources. Wildebeest prefer actively growing grasses, topi prefer intermediate grasses, and kongoni prefer senescent grasses. Each species gains weight most rapidly when feeding on its preferred forage.

What are the advantages of migration? In support of Sam McNaughton's recent work, the short grasses of the volcanic plains are much richer in protein, calcium, and phosphorous than the tall grasses of the northern woodlands. Mineral levels in the northern grasses are so low that nonmigratory grazers would suffer reduced fertility from phosphorous deficiency. Browsers can obtain ample minerals from tree leaves, but grazers must migrate south as soon as the rains permit.

Tony Sinclair: Competition between species can be measured when the abundance of one species is artificially altered. For example, if wildebeest outcompete zebra for food, an increase in the wildebeest population should cause zebra numbers to decline. More food consumption by wildebeest would mean less food for zebra. This "experiment" was actually performed by the elimination of rinderpest. Zebra were not infected by the virus, so their numbers during the epidemic were presumably limited by the availability of food rather than by disease. Although the wildebeest population increased eightfold after the vaccination program, zebra numbers have remained constant. Therefore, the zebra must not be seriously affected by feeding competition with wildebeest or any other ruminant.

A second "experiment" occurred when poachers removed large numbers of buffalo and lions from the northern Serengeti during the '70s and '80s. Topi and impala populations increased dramatically over the same period, either because they faced reduced competition from buffalo or because there were fewer lions to prey on them, or both. Poaching has been sharply curtailed by the recent paramilitary operation, so the precise cause of these changes may soon be clarified.

Ray Hilborn: Sinclair's earlier research showed that rainfall stimulates grass growth and that more calves survive when more fresh green grass is available to each animal. Any rancher could tell you that the size of the wildebeest herd will depend on rain, but now we have models that predict what the *precise* number of wildebeest will be, given recent patterns of rainfall. The current wildebeest population is only about half the theoretical maximum and well below the size expected from recent rainfall. This difference suggests that poachers may be killing as many as seventy thousand animals each year.

Andy Dobson: The Serengeti is at risk from further epidemics of rinderpest. A mild strain of the virus hit Ngorongoro and the northern Serengeti in 1982, but a more virulent strain has recently struck Uganda. Vaccination programs are expensive, and most African countries lack the resources to protect all of their cattle. Because rinderpest can only persist in large populations, it is not always necessary to vaccinate entire herds, except where cattle come into frequent contact with wildlife. By measuring the associations between Masai cattle and wildlife, we will determine the most efficient vaccination scheme for the Serengeti region.

Craig Packer: The Serengeti lion population increased by about fifty percent as the ruminant populations recovered from the rinderpest. Lion population growth was sporadic and depended on annual rainfall patterns; occasional years of favorable weather produced bumper crops of cubs. Lion numbers have increased most dramatically in the woodlands, where resident herds of buffalo are abundant. As the woodland lion population increased, excess subadults spilled onto the plains. Over the same period, the plains prides barely produced enough cubs to maintain the population. The plains provide a marginal existence, and the plains population is largely sustained by immigration from the woodlands.

Heribert Hofer and Marion East: Hyena numbers also have increased dramatically over the past twenty-five years, but their numbers have declined somewhat in the past four. Hyenas specialize on wildebeest and commute regularly to the migratory herds. A female may travel fifty miles before reaching her prey, then spend several days feasting before returning to deliver milk to her cubs. During the rains, hyenas head out to the plains, but they pass through areas of intense poaching during the dry season. In the past two years, ten percent of the animals under study have been killed by poachers' snares.

Tim Caro: Cheetahs specialize on gazelle in the open plains, and the cheetah population has also increased in the past few decades. However, cheetahs are timid and lose many of their kills to other, more aggressive predators. Cheetahs may be suffering from competition with the growing lion and hyena populations. In the past few years, over ninety percent of cheetah cubs have been killed by lions and hyenas.

Holly Dublin: The Serengeti ecosystem extends across the Kenyan border, where it is protected by the Masai Mara Reserve. Like the rest of the Serengeti, the Mara woodlands have been greatly transformed over the course of the century. After the spread of acacia brushlands at the start of the rinderpest epidemic, the Mara became heavily wooded until wildfires removed most of the younger saplings in the '60s. At about the same time, elephants migrated into the ecosystem after agricultural activity intensified in southern Kenya. Remaining within the safety of the reserve, the elephants hastened the demise of the mature trees and further hampered the regeneration of young trees.

Nick Georgiadis: The striking geography of East Africa has created numerous barriers to animal movements, and it is well known that prolonged genetic separation can lead to the evolution of distinctive races. The Rift Valley restricts the movement of several species in northern Tanzania. Wildebeest in the Serengeti and Ngorongoro Crater belong to the same genetic race. But the wildebeest in the Crater and Lake Manyara are as genetically distinct from each other as separate species, even though they live only twelve miles apart. However, the African elephant is not bothered by mere walls or escarpments. Manyara elephants commute to the Crater, for example. The genetic similarity of elephants from Kenya to South Africa implies a long history of large-scale migration.

As I walk back toward our house from the library, my head is still swimming from the morning's discussion. On the dome of a small kopje hyraxes and agama lizards sun themselves on the bare rock, then scramble for cover when a martial eagle soars overhead. A tall acacia shades the ground nearby. The tree may be one hundred years old, but the rock has been there almost forever.

The Serengeti has an extraordinary rhythm. A rhythm of the seasons, of migration, and of life itself. But the hallmark of the Serengeti is its great resilience. The Serengeti rebounds from one disaster after another, emerging transformed from open grassland to impenetrable scrub to picturesque woodland.

Nothing is static here. The animals move around from day to day, the vegetation moves around from decade to decade. Species will change, too—evolution did not stop in the misty dawn of time but remains an ongoing process of modification. The very indomitability of life depends on its extraordinary capacity for endless change.

MONDAY, 9 DECEMBER

Eight men and women sit around a long table in the SRI library. The tabletop is littered with hand-drawn graphs and charts. One person is taking notes in a large spiral notebook. We are meant to be working as a team, but the session has been degenerating. Half the group looks baffled, the rest are irritated and angry.

"This is impossible! It's just a complete waste of time."

Tony and Peter have divided the workshop participants into groups of specialists. Our group represents the major predators: cheetahs, hyenas, leopards, lions, and wild dogs. The completed computer model will partition the ecosystem into different zones (plains, woodlands, Maswa Game Reserve, Masai Mara, and so on). We have been asked to estimate the number of predators in each zone, the number of prey animals that they kill each year, and the extent to which predator populations would rise or fall with changes in the ungulate populations.

We have been sitting together since 10:00 AM; it is already past noon. Participation has not been wholehearted. All of us have spent years pursuing our own specialized interests in our respective study species and we are not used to such grandiose collaboration.

And few of us have studied these broadscale ecological questions in detail; no one wants to stick his neck out. Scientists are in the business of making highly educated guesses; we hate to take stabs in the dark. Furthermore, there is the inevitable rivalry between research projects. Some people have worked on these questions more seriously than others; some species are easier to study. A fair amount of personal pride is on the line.

By the time we break for lunch, there is a feeling of near paralysis. We filter back to our houses dissatisfied and disengaged. However, we quickly discover that the other four specialist groups have all made real progress. Two of the specialist groups are measuring human impact on the Serengeti; one group focusing on the effects of tourism and park development, the other concentrating on agriculturalists and poachers. They will start writing their parts of the master program this afternoon.

The other two groups have already started programming. The vegetation group has estimated grassland abundance, the relationship between rainfall and productivity, and the proportion of the standing crop that is lost to wildfire each year. The ungulate group has assessed fertility and calf survival under different levels of food availability.

Only the predator specialists have been impeded by doubt and dissension, and we are at the center of the whole enterprise. We have to pull ourselves together.

Lunch on the verandah. Sarah, Pam, and Christine have been out looking for lions all morning, but I'm too absorbed in the workshop to contribute to the general conversation.

We have a rough idea of the lions' distribution throughout the entire ecosystem, but we know virtually nothing about their diet in the northern part of the park. Even worse, we don't know how the lion population would rise or fall with fluctuations in each of the prey populations. What if the zebra population were to double? Zebra numbers have always remained constant, so there is no way to know.

Back in the library, a sense of urgency has begun to overtake all caution. We remind each other that we are not designing some new procedure for brain surgery here, we are just providing a rough set of estimates to help park managers decide how best to allocate their precious resources.

Tourists bring revenue to Tanzania. Tourists come to the Serengeti for the wide-screen spectacle, and in particular they come to see the predators—handsome, muscular, red in tooth. How can the park best deliver the product to the visitor? Should park rangers devote more of their time to controlling fire or to catching poachers? Which prey species have the greatest impact on predator numbers? How would management decisions affect predator abundance in the long term?

In the short term, we all stand to benefit from the workshop by learning something about what we don't know. We came here to study the survival and reproduction of these animals. But if we can't specify the factors that limit the size of their populations, we don't really understand their underlying selective pressures. We are scientists; we are supposed to be driven by a spirit of discovery. Here are new worlds awaiting to be conquered.

And then something happens. Everyone suddenly sheds all reserve, rivalry, and caution. Numbers start bouncing around the room. There is a palpable sense of collective concentration, of working as a team. People are talking to each other in a way that has never happened in this room before. None of us believes a word we are saying, and we are assembling the machine with all the wrong parts. But we have decided to put it together anyway; we will sort out the details another time.

Now we can deliver our contribution to Ray Hilborn, stand to one side, watch him bring forth a world that is quite literally of our own making—and see if it bears any resemblance to reality.

TUESDAY, 10 DECEMBER

As the sun rises above the low, rolling landscape, we wind our way through a scrubby stand of acacias. To our left, the narrow concrete houses of the SRI staff village glow in the early morning light. Broken-down vehicles stand like civic sculpture in the courtyard. Early risers in ragged clothes are already walking between open doorways.

"Hold on—what are those?"

Pam stops the car. Well off to our right, at least a dozen golden bodies are moving at a trot toward the brushy cover of the Sangere River.

"Lions—mothers and cubs!"

We drive cautiously. The lions are extremely nervous, startled by the awakening staff workers. We can see four adult females and at least eleven cubs of various sizes. One of the four females is wearing a dark band around her neck, but the transmitter has fallen off.

"Yes! It's the Transect pride, Trixie is still wearing her old collar."

Relocating the Transect pride has been our highest priority for the past week. Trixie lost her transmitter in April, and without radiotelemetry we can go for years without seeing lions in this thick brush.

Pam and Sarah discovered an important lead to the Transects when they found Golfball and Buffalo Bob with the Sangere/Northwest pride last week. The two males had been residents in the Transect pride during the previous year, before they annexed the Sangere pride. We have been tracking the Sangeres every day in hope that Golfball and Bob would eventually lead us back to their earlier family.

Male lions take a utilitarian view of paternity. Rarely adding to the household larder, fathers contribute to family life primarily by protecting their cubs against other males. However, paternal protection is only critical until their cubs are about nine months old. Older cubs can flee from invading males and accompany their mothers to safe havens. The mothers are not ready to start a second batch of cubs until their first cubs are nearly two years old, taking a yearlong break from pregnancy and lactation while devoting themselves to the maturing juveniles.

Instead of sitting around twiddling their thumbs for a whole year, the fathers of yearling cubs will often move into a nearby pride to claim their neighbors' wives. After starting a new family, the males devote most of their energies to protecting their new babies. However, the abandoned wives will shift their range to remain within earshot of their former husbands, who can still be counted on for assistance if the females are assaulted by invading males, and the fickle fathers will occasionally return home to visit their older children.

Although Golfball and Bob are not here today, our persistence has finally paid off. The Transects spent last night within the fringes of the Sangeres' territory. We were fortunate to spot them before they vanished into the scrub.

After traipsing another mile from the staff camp, the pride finally settles on a low, flat rock beside the tree-lined riverbed. Though it rises only a few inches off the ground, the rock somehow gives the lions the same sense of security as a massive kopje.

We spend the next two hours drawing spots, updating the lions' ID cards, and waiting for everyone to calm down. The oldest cubs are quite curious. They are examining the car, walking around us, and staring in through the windows, but they are soon distracted by better games—carrying sticks and batting the twitching tips of their mothers' tails.

Except for Trixie, the mothers have gone to sleep. Trixie's cubs are the youngest in the batch and are not yet used to cars. Responsive to their anxieties, Trixie watches us alertly as we prepare a dart.

"Okay, Pam, fill the barrel of the syringe with anesthetic and attach the needle to the barrel. Press it on hard with the pliers. If the needle's not on tight, it will pop off before the drug is injected. Make sure that the collar covers the holes near the tip of the needle, otherwise the drug will spray all over you. Now inject the butane through the silicone plug at the other end of the barrel. Check that there are no leaks, and then screw the tail-piece in behind the plug. Well done!"

The finished dart is a plastic syringe fitted with a red-finned tailpiece. The collar slides back when the needle penetrates the animal, the butane pushes down the plunger, injecting the drug with the speed of an enthusiastic vet.

The farther the shot, the more likely something will go wrong, so we try to get as close to Trixie as we can. Although the lions are settled for the moment, one false shot now and the Transects might disappear down the riverbed. This is our only chance, and we need a clear shot from within fifty feet.

Pam starts the car and drives slowly and cautiously. But Trixie gets up quickly, moves around the far end of the rock, and lies down beneath a bush.

We wait a few minutes for Trixie to relax, then drive in a zigzag pattern. Just as Pam makes it within range, Trixie stands up and walks off.

We sit frozen in the car. Trixie settles again, about seventy feet away.

After five minutes, Pam starts the engine, drives forward a few yards.

Trixie gets up and walks out of view behind the bush.

Stop again, wait five more minutes. The sun is starting to beat down. The rest of the pride has gone into deep flop.

Trixie finally stretches out in the grass. We move to within seventy feet; Trixie jerks her head up as if she wants to move off.

Dead silence in the car. Trixie puts her head back down.

Five more minutes.

Move in ever so gently to forty feet, slow to a smooth stop.

Breaths held for immeasurable seconds.

I fire and hit her square on the shoulder. We can see that the plunger has been pushed all the way in. The drug has been delivered.

Trixie gets up and walks thirty feet to the rest of the pride. Several cubs come up and rub their heads against her legs. Unperturbed by the injection, she lies on her front and pulls the dart out with her teeth. She examines it with her tongue and mouth for a minute or so. One of her cubs comes over and takes it away from her. She lies back down to resume her morning nap.

The tension has broken in the car, and everyone breathes normally again.

Seven minutes after the injection, Trixie is well under the influence. Pam drives up to her, stopping with Trixie just beside the passenger door and the other fourteen lions resting ten yards to the front of the car.

Sarah carefully opens the door and props it wide-open to block the other lions' view. She slowly slides out and removes Trixie's old collar. From the safety of their rock, the rest of the pride calmly watch Sarah's disembodied arms and feet below the open door. Sarah works quietly—not for fear of her own safety, but to avoid frightening the rest of the pride. If they realized what was going on, they would panic and flee and become even more frightened of people in the future. But as long as Sarah remains within the contour of the car, she is not recognizably human.

The old collar is off, and Sarah uses it to measure the correct length for the new one. Pam turns on the receiver to make sure that the new transmitter is working. After trimming the new collar, Sarah places it around Trixie's neck and tightens the bolts.

Mission accomplished.

Pam slides out to help Sarah roll Trixie onto her side. They measure her girth—40 inches, slightly larger than average for a Serengeti female, about 250 pounds. Pam lifts Trixie's hind leg, and Sarah has the magic touch; blood quickly fills the syringe.

Sarah and Pam stroke Trixie's side and take one last long look. Breathing easily and lying comfortably, the lion is the object of their compassion rather than their fear. They slide back inside and Pam backs the car ten yards away. The rest of the Transects calmly watch their sleeping companion for a minute or so, then resume their own slumbers.

Trixie starts to come round, and once we know she has recovered, we head back to SRI, drained and exhausted. But my own exhaustion is mixed with elation, not only at our success, but also that Pam and Sarah's enthusiasm is balanced by such obvious sensitivity.

WEDNESDAY, 11 DECEMBER

Back at the symposium, anxious clusters of researchers stare intently into their computers. Each specialist group is rushing to complete its assignment. The day has dawned dull and cloudy, and the solar panels can barely generate enough power to meet the demand. People dart from one room to the next.

"How are you measuring grass production in the north?"

"How many villages in the Maswa?"

"Where are the wildebeest in October?"

The vegetation, herbivore, and park management groups deliver their completed programs to Ray Hilborn.

The solar batteries are starting to die; there is only enough current to run two computers.

"Cut cub survival in half when the population reaches seven thousand."

"Say that goats spread the virus to cattle but not to wildebeest."

Programs from the carnivore and regional development groups arrive next. Finally, miraculously, the master program is ready to go, and we all assemble in the mapping room to watch the computer screen.

Underlying everything is rainfall. More rain equals more grass. More grass equals more grazers. More meat equals more predators.

Ray plays god, instructing the computer to reproduce the sequence of rains that have fallen on the Serengeti in the past thirty years. A quarter million simulated wildebeest should increase eightfold by 1990; the simulated zebra should stay steady at two hundred thousand.

The computer starts its calculations. The lines on the graph inch their way across the screen. The wildebeest population increases as expected, but the zebra soar to three million.

Tinker, tinker.

Six hundred cheetah skyrocket to one hundred and twenty thousand.

Tinker some more.

Lions become extinct because hyenas eat all the wildebeest; cheetah become extinct because lions eat all their cubs.

Estimates are reappraised, guesses fudged until population numbers rise and fall in approximately the right way. The simulated Serengeti eventually shows the same changes in richness and diversity that we had actually measured in the field. Either we have finally tweaked all our guesses to a reasonable approximation of reality, or our innumerable wrongs have somehow made a right.

Ray awards himself another thirty-year term in office, allowing the rain fall between 1990 and 2020 to follow the same sequence of drought and flood as in the previous three decades. Unlikely, to be sure, but our goal is to assess the long-term consequences of management policy, not to make a killing on the stock market.

How can park managers best cope with the quarter million people that live along the western edge of the park? The human population is increasing at three and a half percent per year—a doubling time of only twenty years.

Ray and Tony consider several scenarios. First, pretend that the paramilitary operation of 1990 never happened and allow poaching to increase in direct proportion to the human population. Wildebeest and zebra are driven to extinction by 2020; lions and hyenas are virtually eliminated. Next, terminate all antipoaching effort. Pretend that the park rangers leave their posts this afternoon and go back to their villages. The extermination is brought forward to the year 2000.

A new rinderpest epidemic or a long-term drought, coupled with even moderate poaching, also leads to mass extinction within ten years. With so many more human mouths to feed, a diminished wildebeest population would quickly be annihilated.

However, two scenarios suggest a more hopeful future. If the intensive effort of 1990 can be sustained for the next thirty years, and a new rinderpest epidemic can be avoided, the wildebeest could increase to two and a half million within ten years, and most of the predator populations would increase along with them. A similar boom would result from the AIDS-related deaths of five percent of the human population each year.

If Tanzania follows the "Fortress Serengeti" strategy, the draconian measures might be enough to maintain the abundant herds. But history tells us that such repressive measures will not be tolerated forever. The poachers are Wakuria, Sukuma, and Wa-ikoma. The rangers come from all over the country. The Wakuria have raised one leopard-skin flag up the pole already; memories are long, and time is short.

Humanitarian considerations aside, there is no point in assuming that the AIDS epidemic is going to solve anyone's problems. Removing five percent of the adult population each year would lead to a breakdown in society. A nation of orphans would be left to survive by whatever desperate means possible.

The regional development specialists are quick to point out that the Serengeti's future could otherwise be secured by reducing human impact by only five percent a year, and they are full of ideas for diverting energies to more positive ends. Poachers are naturalists who have to appreciate animal behavior to capture their prey. Perhaps they could become tour guides in the game reserves, their families employed by new lodges. Livelihoods would depend on protecting resources, rather than destroying them. Reward schemes for collecting weapons and snares could be expanded; meat hunting could be legalized and regulated, bringing the trade out into the open. Finally, development could be accelerated at a safe distance from the Serengeti to draw people away from the free meals at the park boundaries.

Although most of the delegates here may accept the large-scale implications of the computer model, many are openly skeptical about these projects. Resources as diffuse as "wildlife" are rarely protected for long. Animals wander from one homestead to another, and no one herd is the responsibility of any one person, family, or village. The situation leads to a "tragedy of the commons," the temptation to destroy any animal that wanders into your own backyard while admonishing your neighbor to protect whatever wanders into his. The only solution is constant law enforcement.

Should the value of wildlife even be expressed in economic terms? Intensive agriculture can produce far more revenue per acre than any wildlife utilization scheme. Most Tanzanians respect the intrinsic value of their natural resources. Why put a bounty on every head? And, anyway, why should Tanzania carry the entire burden of protecting the Serengeti?

Sitting beside the Loliondo pride at dusk, all the dire predictions seem a million miles away. The lions are snoozing, the grass is brilliant green, and countless hooves will be pounding through here in another week or so. The circle must surely be unbreakable.

The golden sunset underscores the fleeting urgency that evoked such a remarkable display of cooperation. For a few days at least, the fractious, divided community was molded into a coherent whole. I almost wish I weren't leaving so soon.

On the other hand, I don't know if I could wait much longer for Erin to hand over those samples.

THURSDAY, 12 DECEMBER

Solo breakfast on the verandah. The view is shimmering and fresh. Sunbirds twitter in the morning light, bronzy-winged, and scarlet-chested. Glittering green mariquas flutter between bursting flowers. Crystalline beads of dew are still in the low grass. Glowing amber boulders frame the distant hills. The hills could be a hundred yards away, they could be on the moon. Air is never this clear, so who is to say?

Pam and Sarah come back, frustrated. The Land Rover won't start.

The view is still there, but the moment is gone.

Open the bonnet, turn the key. Click. The connection to the starter motor is okay. The charge light has been on all week, so maybe the battery is dead. Switch on the headlights and turn the key. The lights dim to a dull orange.

Walk to the Lion House to fetch a spare battery, lug it back in time to find Rob connecting jumper cables from the Toyota to the Land Rover. Setting the battery down, I lose my grip, drop it three inches, and crack the case. Acid leaks out over the roadbed. Pam drives the car to the Seronera garage.

I stroll back to the vault for a final inventory. Do we need more radio collars? Antennas? Inner tubes? Tools? Will one more car battery be enough? Don't forget new wiring for the Suzuki, a new hose to fill the hot-water barrel, buckets to haul water.

People drop by to say farewell. The Olympics are over, but there is no lowering of the flag, no singing of the anthem, no dousing of the flame. Tony Sinclair feels flat. The workshop has been a success, but he had to pull so many teeth to bring it off that he is thinking of giving up. He has enough to do in the Yukon and Australia as it is. He doesn't need another headache.

"You can't quit now, Tony. You are the one who finally convinced me that it was worth carrying on out here. Until a few days ago, I felt like there were no mysteries left to the Serengeti. But you've given me a whole new set of problems to solve. Everyone has started looking at their work in a new way. You've had an impact on all of us."

"You really think so?"

"Didn't you see all those people talking to each other? You've got to come back."

Then farewell. But he will be back. And he doesn't need any pep talks from me—we all come back. The Serengeti is in our blood.

I am beginning to feel organized. Sarah has transcribed all her data, and Christine has nearly packed. Pam comes back from the garage. The wire to the starter motor was shorting out against the chassis, the battery is fine.

We sit together on the verandah for our last Serengeti lunch. The sun is straight overhead, and the sky is a brilliant blue. The acacias have been frosted with white flowers, hyrax mow the lawn, and at our feet the birds have started molting into their breeding plumage.

Startled, the birds chirp loudly and flap up into the trees. A pack of dwarf mongooses has arrived, seeking bits of cheese and bread. Their reddish-brown fur is mottled with white patches where they have been freeze-branded for permanent identification.

A dwarf-mongoose pack contains up to a dozen animals, but breeding is restricted to a dominant pair. Before they reach breeding status, youngsters must wait in a "queue," and until they reach the top of the ladder, subordinates act as helpers to the breeding pair. A subordinate female may even start lactating to help nurse her mother's litter, but her own reproduction is suppressed by her mother's constant harassment. Kindly subordinates or manipulative dominants, depending on your point of view.

Dwarf mongooses may be fascinating, but the mere sight of them is enough to dispel any warm glow from the past few days. The Serengeti mongoose study was established in the 1970s by the late Jon Rood. Jon's discoveries revealed a complex social world scurrying half-hidden in the grass, searching for insects, stashing their babies in termite mounds, screeching at the first sign of danger.

Jon was one of the most respected fieldworkers of his generation, but he suffered from clinical depression, and his self-esteem could be sent unpredictably into a downward spiral. He spent months each year in the US, writing up his observations and publishing papers in scholarly journals. Scientific manuscripts are reviewed anonymously by colleagues who recommend changes and advise an editor whether to accept the paper for publication. These peer reviews are meant to be impartial and constructive. Most referees do behave responsibly, but competition pervades science as much as any other human endeavor, and some will exploit their anonymity by trying to thwart the efforts of their rivals.

We all have to endure the harsher criticisms and the occasional dose of poison, but Jon was particularly fragile the day he received two rejection notices. In one of the critiques, hidden artillery had delivered a direct hit. The hostile review triggered a major depression, and Jon drowned himself two weeks later.

Late afternoon in the Toyota, Rob and I are driving along the Seronera River looking for CSN, the female that eluded us on our first night in the Serengeti. She has kept herself hidden in the same den for weeks. I want to see her cubs before I go, and Rob would like to find her with the rest of her pride. CSN's sisters have formed a large crèche with a dozen cubs and subadults. Rob has brought along the equipment to perform a playback experiment.

Rob is focusing on the development of cooperative territorial defense in young lions. Using the recorded roar of a female from another pride, he has found that young females are quick to join their mothers in expelling the apparent invader. The young males, on the other hand, hang back and let the females resolve their own differences.

CSN is out in the open today, but she is no longer lactating, and she is not with the rest of her family. There has been a takeover. Her former mates have been ousted by Buffalo Bob and Golfball. Bob is lying a few feet away, eager to introduce himself. CSN must have been hiding all this time to give her cubs a small chance of survival. But her litter has been lost, and it is time for her to start over.

Lions are not expressive animals, but CSN almost looks relieved.

We drive back home to SRI and find that the tension there has finally broken as well. Erin has delivered the hostage fecal samples to Christine.

FRIDAY, 13 DECEMBER

Good-bye to Erin and Rob, and away we go. Down the wooded lanes, past Seronera and Banagi Hill, on into the north. The sparkling, shining day unfolds through an endless Eden of impala and giraffe, all orange and brown against lacy green trees. Klipspringer tiptoe on towering grey kopjes, and lines of wildebeest funnel south toward the open plains. Eland bound across the road. Buffalo and oribi, dik-dik and gazelle, topi and kongoni, on and on and on we roll.

Forty miles north of Seronera, a dazed, dusty man walks uncertainly beside the road. He stares vacantly into space, his hair has gone spiky, his clothes are in rags. Twenty miles from the nearest village, he is a head-case gone walkabout. An oncoming car full of park rangers stops him for questioning. Sarah and Pam are upset by the thought of his dark nights, alone and disoriented, in the middle of paradise.

We pass Lobo Lodge, blended into the outlines of a giant kopje, and enter the danger zone. Virtually abandoned when the Kenyan border was closed, the road runs through rugged country dominated by high-crested hills, dark kopjes, and deep swollen rivers. Over there is the spot where the German tourists began their barefoot journey.

Now the view sweeps away to the west, down where so many elephant roamed before the poachers discovered the value of ivory. The surviving herds sought refuge in the Masai Mara, leaving behind the broken trees, the occasional white skull gleaming in the sun.

Next Bolagonja, the Tanzanian border post and source of the spring water that was once piped to Seronera. Our passports are checked, the car papers signed. The customs officer wears a T-shirt that says "Redskins beat Dallas." Everyone is so very pleased that Pam and Sarah will be returning soon. Officials, policemen, and park rangers all hand over their shopping lists and wish us safe passage.

One of these rangers once told me, "The Kenyan border had to be closed. Capitalism next to socialism is like fire next to water." President Nyerere attempted to construct a socialist society that was based on the traditional values of the extended family and the local village, but these small-scale concepts failed miserably at the national level. Although no one prospered in socialist Tanzania—"He didn't want anyone to get rich, so he made sure that everyone stayed poor"—neither did anyone feel left out. As disillusioned and corrupt as this country has become, there is still a sense that each person has something to contribute to the community, that everyone belongs. But there is everywhere this awful temptation to get rich quick, to take the bribe, to sell out in a hurry.

Tanzania's naive social policies have left the country in a position to learn from the mistakes of others, to maintain its sense of community and develop itself from scratch in a sensible manner. But since no other country has ever managed this feat, it seems hopelessly optimistic to expect Tanzania to be the first.

We drive ten nerve-wracking miles to the Kenyan border, through the real no-man's-land where bandits hide from the Kenyan police between attacks on rich tourists in the Mara, where one of my assistants was once pelted by stones, and where an Englishwoman was murdered by Kenyan rangers. But the only excitement today is a large tortoise waddling along the road. We stop to carry him into the safety of the tall grass.

At the Kenyan border post, the police inspector insists on opening our luggage. He rummages through Christine's bag of shit samples, and he is not at all amused when we tell him what the vials contain. After fifteen minutes of hostile interrogation, he rudely orders us into his office. The inner walls have been painted blue and orange. A photograph of President Moi hangs next to the xeroxed portrait of an unrecognizable African military chief. The handwritten caption reads: "Army Boss."

While the policeman laboriously copies the details of our passports, I point out a wall calendar that advertises Nakumatt Supermarket in Nairobi. A large portrait of Jesus, Mary, and a variety of angels appears above the slogan “Nakumatt: much more than a supermarket.” My companions laugh too loudly and make derisive comments within a few feet of our host. Fortunately, the policeman is too engrossed to notice.

This is not a man to humiliate. He has no sense of humor and more power than I like to contemplate. But as we return to the car the reason for his intimidating behavior finally becomes clear. He wants us to give his friend a lift to town. The friend hops aboard, and the now-smiling policeman waves us on our way.

At Keekorok Lodge, we stretch our legs, beat the dust from our faded shorts, walk in past the reception desk, and, Toto, something tells me we’re not in Kansas anymore.

The lodge is full of jaded Western tourists who lounge around low tables in overstuffed chairs. Here in the wonderland of Africa, the main attractions are the extravagant buffet and the elaborate cocktails. No one seems to notice the purple bougainvillea or the orange trumpet flowers. No one watches the buffalo coming down to drink at the waterhole or the vervet monkeys in the fever trees. We stumble past the bored teenagers reading magazines, the husbands strutting in safari suits, the long-lashed glances of their wives.

Marching past the par, we find Holly Dublin, just back from the Serengeti symposium. Holly introduces us to Joseph, a Kenyan graduate student who has been studying the lions in the Masai Mara Reserve. Far smaller than the Serengeti, the Masai Mara is much better protected. Lions multiply rapidly inside the reserve, and the dispersing subadults are siphoned off to the surrounding cattle lands where they are quickly eradicated by Masai herdsmen. Joseph and Holly are worried that the Masai might try to solve this problem at its source. Poisoning is already common outside the Mara, but what if people start putting poisoned bait in the middle of the reserve? Not long ago, the lions in Amboseli National Park were exterminated when the Masai laced a dead donkey with sheep-dip and killed the only pride in the park.

The Mara may be small, but it holds many more wildlife lodges than the Serengeti. Kenya has developed far more rapidly than Tanzania, and most wealthy developers transport their riches to Nairobi, leaving local herdsmen to cope with the wildlife that spills out from the park. If the disgruntled locals eventually destroy the animals, that's okay too. The developers can always sell off their land for agriculture. A buck is a buck.

Walking outside, we meet the young assistant manager of Keekorok Lodge. Like Joseph, she is a middle-class Kenyan and well-educated, the product of a functioning economy. She stops us for a moment to ask about our research, about lions, and about the future of all the other species in the Serengeti. She asks not from courtesy or because it is good for business, she asks because she cares. "At night, when I hear the lion roar and the hyena whoop," she says, "I feel safe."

Passing through the park gate, we finally leave the protected lands. We have traveled two hundred miles from the entrance of the Ngorongoro Conservation Area, one hundred miles from Seronera. Here, the Masai herd their cattle in every direction, mud huts dot the grey-green plain. After another half hour, we pass a nondescript spot on a nondescript stretch of road, the spot where I once rolled one of our cars.

I was driving back from Nairobi with a friend from SRI. We were so heavily laden that we could only travel about thirty miles an hour. A tire blew, and slowly, ever so slowly, the car swerved off the road, slid sideways into a ditch, and tipped onto its side. But the load was so great—timber, bags of cement, car spares, buckets of white paint, a large supply of food—that the momentum carried the car onto its roof.

Hanging suspended by our safety straps, we assured ourselves that we were both okay, unbuckled, and landed headfirst in a puddle of paint on the ceiling. My companion's hair instantly turned white. He caught a lift to town with a minibus full of horrified tourists while I stayed behind to guard the wreckage.

Two old Masai men appeared from nowhere. They looked at me with an expression of compassion mingled with disdain. One muttered something to the effect that, well, what do you expect from these crazy *wazungu* with all their gizmos and machinery?

Although the damage was costly, I was most affected by the banality of the danger: thirty miles an hour on a flat stretch of road. My view of raising a family in Africa was forever changed. It was like that terrifying moment when you first discover that any table, chair, or toilet bowl can knock out your kid's teeth—you live with it but every now and then it scares you to death.

The following year, Anne and I returned to the Serengeti for our last long-term visit. Fieldwork no longer gave me any joy. Lions were inanimate, and life was suddenly too short.

The battered car sighs with relief as we finally reach the pavement. We pass slowly through Narok, a boom town of tawdry tourist facilities. The main street is lined with gas stations and tourist kiosks decorated with crudely painted cut-out animals.

We dip up and down over seven rollercoaster hills before reaching the top of the rift escarpment that stretches far, far below. On a clear day we would be able to see Oldoinyo Lengai and the rift walls almost to Manyara, one last glimpse of Tanzania. Today, though, it is too hazy and we can barely make out Suswa and Longonot, the jagged Kenyan volcanoes that lie ahead.

Climbing up the crumbling escarpment road on the far side, we pass the spot where a middle-aged missionary couple was hijacked one night last year. The husband was forced to watch while his wife was raped to death.

Barreling through the caffeine plantations of the fabled Kenyan highlands of British decorum and white mischief, we enter the comfortable Nairobi suburbs and arrive at Barbie Allen's house in time for tea.

We receive warm greetings, a dusty hug, and introductions to various old friends on the verandah. Cups filled, conversation turns to the recent attack on an elderly white settler in Nakuru. Badly hacked by a *panga*, he has lost an ear, an eye, and nose.

Land must be distributed to Africans upon the death of a white owner in Kenya. This country is land starved, the population is exploding, and "They always go for the elderly when they want land."

NAIROBI / SATURDAY, 14 DECEMBER

Walking along the crowded sidewalks in the fierce glare of the sun, I'm spending my last day with Pam and Sarah, the final day of class.

"This bank stays open till four. Go to Moi Avenue for car spares; have the films printed on Kaunda Street. Electronics can be repaired on Muindi

Mbingu." Food here, kitchenware there. Hardware, stationery, books, groceries, and shoes. Vet supplies, pharmaceuticals, the best place for souvenirs.

Nairobi is the Geneva of black Africa. UNEP, UNDP, IUCN, FAO all have headquarters here. Nairobi has a higher proportion of Mercedes than any other city in the world. Two-year tourists and fat-cat African executives are known collectively as the *Wa-benzi.*

You can eat Japanese at the Akasaka, Lebanese at the Red Bull, Cordon Bleu at the French Cultural Centre. Business is booming, and new highrises soar above every downtown block, but the streets are lined with beggars and the homeless, drunks and prostitutes, cripples and lepers. They lie dazed beneath neon lights, evidence of the human tidal wave swelling out of the shantytowns on the city fringes, latent statistics in the raging AIDS epidemic. The glitter in shop windows is locked behind burglar bars and heavy iron chains, protected against the next round of riots, the constant threat of theft.

Kenya is dominated by three or four tribes. Her first president, Jomo Kenyatta, ensured that his own tribe prospered at the expense of his rivals. As political pressures mounted and Kenyatta aged, everyone predicted civil war. But when Kenyatta died, he was quietly succeeded by the vice president, Daniel arap Moi. Moi was mild and pragmatic and came from a tribe too small to affect the balance of power. The succession inspired the happy headline, "*Après le Déluge*—Moi!"

But Moi's small tribe quickly formed a confederation with its neighbors to withstand the larger tribes, and Moi has become one of the richest men in the world. He vigorously defends single-party rule, claiming that in a multiparty state, tribalism would tear Kenya apart. Government critics disappear mysteriously, newspapers are heavily censored. Mobs erupt in the streets, and order is brutally restored. Protected by his pro-Western stance during the Cold War, Moi has only recently come under pressure from the international community. Two weeks ago, he finally permitted the formation of an opposition party.

"This is Biashara Street—Swahili for 'bazaar'—the best place to find cloth and bedding." The narrow street is lined with small shops: Bazaar Emporium, Atul's Tailors, and Kenyan Drapers. We pick a store at random;

everyone wants to buy African cloth to send home for Christmas. The single-aisle shop is decked from floor to ceiling with bold patterns, all the bright colors that vibrate so intensely in the clothing of African women. The indoor air carries a faint trace of incense. The salesmen and craftsmen are African, but the shop is owned and managed by Asians.

The African salesman seems confused by Sarah's order, and the owner, or his brother or cousin, quickly steps from the dim recesses to complete the transaction. Everyone eventually finds something to buy, and the Asian snaps at his salesman to write up the bill. A second Asian stands idly in the back room, practicing his golf swing.

Thrown out of Uganda and widely resented in both Kenya and Tanzania, the Asians prosper nevertheless. They refuse to assimilate into local communities, and their shops are always the first to be looted during a riot. Many African countries discriminate against Asians, preventing them from rising too high in any large corporation and restricting their access to higher education. But the Asians are essential to local commerce and somehow manage to endure. Maybe they are protected by an entrepreneurial spirit that generates jobs without denying anyone access to the land. Preadapted to city life, able, like my old friend Rasul, to find business at the drop of a hat.

We meet Barbie for lunch at an Indian restaurant in the middle of town. Job training has ended at last, and I can finally start to relax. The conversation is lively, and Barbie and I resume an ongoing dialogue, swapping jokes and recent news.

But then Barbie announces that she is trying to sell her dry-cleaning business. Most of her children have emigrated to Australia, and life in Kenya is increasingly difficult. Her equipment must be replaced every six or eight years. Import duties and business permits have become more trouble than they are worth. I ask if she wants to emigrate, but then she surprises me, as she always does, with her commitment to the community. She doesn't have any plans to leave Kenya, she just wants to work at home full-time and devote herself to civic committees and worthy causes in Nairobi.

Back at Barbie's house, Markus Borner has just arrived from the Serengeti. Markus originally came out from the Frankfurt Zoological Society to carry Grzimek's toothbrush, as he likes to say. But he developed the Serengeti ecological monitoring program so successfully that he has recently assumed far broader responsibilities. He currently serves on the boards of trustees for Tanzania's National Parks, the Ngorongoro Conservation Authority, and Mweka Wildlife College.

We spend the afternoon over a pot of tea on the verandah. Markus is astonished by the rapid changes in Tanzania and sees dozens of new opportunities for research and wildlife conservation throughout the country. New land is still being set aside for formal protection—even through the initiatives of local villagers in some cases. Poaching may be terrible but at least the reserves will still be there once the country gets on its feet again.

Markus thinks it might even be possible to rebuild SRI someday. But he is only one person, and Frankfurt does not have unlimited funds. The Tanzanian government barely has the resources to pay its few remaining staff, and the newfound community spirit among the expatriate researchers has already started to unravel.

Markus is undaunted, but I'm ready to leave.

Standing beside the red Land Rover, I go over plans with Pam and Sarah to pay for Kisiri's cows if he is acquitted, priorities for finding lions in the next few weeks, car parts to be mended. "And never, *never* drive through the Mara if you can't go in convoy. Write often and call me whenever you come to town." Then farewell.

But this is no ending. They are eager to start on their own.

SUNDAY, 15 DECEMBER

Rome. Christine and I rush into the terminal, hurrying to make our connection to London. Brightly lit signs direct us down the stark square walkways. Then we panic. All incoming passengers must pass through a portal labeled "Sanitary Control." Christine's bag full of fecal samples may help clear our way through the crowded tunnel, but it certainly doesn't qualify as sanitary.

Reaching the gate, we find that the "control" consists of a uniformed man distributing small yellow cards. "You have been to a tropical country.

Should you become ill in the next two weeks, show this card to your physician."

Supper on board Alitalia comes complete with two small bottles of champagne. We are ready for the tastes of civilization, but any feelings of relief and satisfaction are overwhelmed by a sense of letdown. Christine is depressed at going back to Oxford, where no one will fully appreciate what she has been through.

I try to warn her about *le mal d'Afrique.*

"Africa is an addiction. Life there is so challenging that you feel you've done something really fine—really rewarding—just by surviving from one day to the next. There is so much that is new, exotic, and exciting, you feel like you are discovering unknown secrets to a lost world. But the last few weeks have just been a glorified adventure holiday. It is what you extract from all that raw material and make your own—that's what matters, that's your contribution."

Cold words to someone who is about to stare at shit for the next two years, but I have seen too many people drop everything to stay on in the warmth and radiant sunshine, hanging around Nairobi, thinking they are doing something grand. The years trickle by, and they might as well have been at the movies.

To me, the real excitement comes from the research, from figuring out the puzzle, working out the rest of the story. Take the most unpromising material and learn something new about the world around you. As creative as art, as challenging as a murder mystery, and a team effort that transcends the sum of its parts.

With luggage in hand and nothing to declare, we march through the gates of Heathrow Terminal Two. A surprise. Anthony has come to meet us, bringing the data—the big charts that I had left at Jane's house during my amebic haze. Our three-way reunion is brief, and we have to rush out to the cold rainy streets to send Christine on the bus to Oxford. Her hard slog is about to begin in earnest.

I check in to an ultramodern airport hotel. The endless hallways are lined with trays of half-eaten meals. The air reeks of cigarettes and industrial carpet cleaner. The room is dominated by outsized beds, synthetic drapery, and a sparkling white bathroom. This is not the Milimani Fisheries Resort. This is a room that demands some chaos.

Anthony and I discuss future plans for Gombe and arrange our next meeting in Minneapolis. He darts off to catch the last tube to London, and I open my bag to find a leaking fecal vial.

Oh, for an open window.

LONDON / MONDAY, 16 DECEMBER

I awaken to a peculiar darkness at 6:00 AM, still two hours before dawn. Might as well look for something to eat. A hotel breakfast would cost a bomb; lights are flashing outside the window. I open the drapes to see—yikes!—McDonald's. Who could resist? Total immersion into the horror, the horror.

The militantly cheerful decor stands defiant against the predawn gloom. Christmas Muzak makes up for the long escape from enforced holiday cheer. The queue moves briskly, and the stainless-steel kitchen delivers a styrofoam package to my vinyl tray. The sanitized "McBreakfast" evokes a sudden, sharp pang for my lost vegetarianism.

Where is the rosemary now? Where are the three young ladies?

All those vivid African smells, colors, and images won't leave me alone. Not for a long time. The warmth, the companionship, the variety and excitement, the cheap thrills of day-to-day life in the bush.

Sitting at my plastic table, blasted by plastic music, I tell myself that I'll get over it soon enough. I've always gotten over it before. I continue my quest wherever I go; it is all in my head. I don't need physical adventure.

Not me.

In Heathrow Terminal Three, the public address system announces that all public transport into London has been temporarily suspended by terrorist activity. The IRA wishes you all a very Merry Christmas.

Crammed on board yet another airplane, somewhere over the Atlantic, I crack the blind briefly to see the low winter sun shining timidly above dull grey ocean. The world is turning, but we remain stationary in space. Time has disappeared.

A movie flickers on the screen at the front of the cabin. In the opposite aisle, a passenger chats hopefully with a stewardess, who manages to extricate herself without creating a scene.

A family sits two rows in front. The mother tries to impose order on her brood. The youngest wants to run down the aisles, kick the seat in front of her, drink another glass of water, raise a little hell. The father reads a newspaper and glances angrily at his harried wife.

I sit surrounded by strangers for the first time in months. I feel as alienated as the moment when I first walked into Keekorok Lodge. I can't help but watch my fellow passengers as if they were just another species preoccupied with the same old questions of sex, violence, country, and the well-being of their children.

But it is abundantly clear that humankind is fundamentally different from any other species. We are animals, to be sure, but we order our societies on relationships *outside* our small family units. Fill this plane with lions, chimps, or baboons, and all hell really would break loose. These humans, on the other hand, are as orderly as ants.

But the essence of being human has nothing in common with ants. We may live in concrete nests piled on top of each other, we may file in and out of our planes and freeways in neat lines, but we are making it all up as we go along. An ant is born into a complex chemical environment where every small instruction has been laid down in advance. Mother tells the workers what to do and they do everything for the greater good of their enormous family.

In contrast, every human being is capable of working for the advancement of their own procreation, their own minuscule families. Yet we somehow recognize the value of a larger form of society, and readily respond to a larger world beyond our own narrow self-interests. With our unique creative capacity, we have modified ourselves as we have modified our physical conditions, and we have developed an extraordinary division of labor. You and I may be as different as night and day, but that is our strength, and it is precisely this diversification that makes my time in Africa so intensely satisfying. My life out there is full of working friendships, necessary neighbors, pulling each other up from the chaos, casting searchlights into the unknown.

But wherever we live, the fundamental currents of competition and ambition still flow through the heart of our deepest desires. We keep up the appearance of working toward a set of lofty goals and seldom stop to ask why we never learn from history, why we condemn ourselves to repeat the same mistakes over and over again. Our failures will never cease unless we know something about our true underlying nature, the nature that doesn't change when a government declares itself capitalist or socialist.

We set such high standards for each other that we find ourselves constantly disappointed, and we become increasingly cynical as we grow older. But to claim that humankind is uniquely vile belies a tremendous naïveté about animals that is as misplaced as the most bizarre form of religion. Animals focus narrowly on short-term selfish gains, on instant gratification, on exploitation and fierce competition.

By studying cooperation in other species, I have watched animals at their kindest. But what have I seen? A handful of helpers exploited by their elders, a few individuals cooperating to make war. The darkness is the behavior of animals. This is what Conrad feared. He placed it in the Congo, but the darkness lies coiled in all our hearts.

I study the darkness. I think about it all the time. I would like to believe that by understanding the nature of selfishness we may someday understand the best way to divert that selfish energy to the common good. Comrades in arms may have been the stuff of romantic images in the past, but it is time to abandon Agincourt for Brussels and convert our tribal energies to nobler ends.

Marcel Proust once said that the most extraordinary journey would be to see the same familiar places through the eyes of another person. But what if we could see ourselves through the eyes of another species? After watching animals, we can't help but see something remarkable, something fine, however disappointed we may all feel in each other most of the time. Look through the eyes of another species and perhaps the occasional good we do will stand out from the constant stream of wreckage.

Postscript

MINNEAPOLIS / TUESDAY, 12 APRIL 1994

On 3 February 1994, a nomadic male lion was found near the Seronera River suffering from *grand mal* seizures. He died during the night. During the next two months, five other Serengeti lions were seen dying with the same symptoms, and dozens more simply disappeared. The Campsite, Loliondo, Plains, Simba, and Transect prides have been hit the hardest—some may have been reduced by eighty percent. The prides at Barafu, Gol, and Naabi have so far been spared, as have the Crater lions.

Many lions now show signs of neurological damage. Some have a twitch that contorts half their face into an involuntary sneer, others jerk and flail their paws whenever they relax. Perhaps they have survived the worst, or perhaps their infection will eventually intensify.

Pam and Sarah have kept me informed by fax. They have been helping the new National Parks vet, Melody Roelke, collect blood and tissue samples, but they have been overwhelmed by the scope of the epidemic, and the cause of the disease remains unknown.

Three days ago, a group of six men attacked SRI. Armed with *pangas*, spears, and poisoned arrows, the invaders bound and gagged a family of researchers. Kicking and hitting the father with a *panga*, they threatened him with death and demanded dollars.

Pam and Sarah report that about twenty percent of our study lions have died so far and that a team of Tanzanian vets will be arriving in a few days.

I leave for the Serengeti tomorrow.

Acknowledgments

By having been granted the opportunity to write about changing times in faraway places, I have benefited from more than my fair share of goodwill. My family has shown exceptional patience, although my children did insist on growing two years older in the meantime. My wife, parents, and colleagues at the University of Minnesota have all been remarkably tolerant of my incessant travels. I suspect, though, that they probably need a rest from me every now and then.

The lion research conducted over the past quarter century by George Schaller, Brian Bertram, David Bygott, Jeannette Hanby, Anne Pusey, and myself has been supported over the years by the New York Zoological Society, the Royal Society of London, the Harry Frank Guggenheim Foundation, the Eppley Foundation, the National Geographic Society, the American Philosophical Society, Sigma Xi, the Graduate School of the University of Minnesota, and the National Science Foundation. The inspiration for this book arose during a sabbatical at Oxford University, where I was supported by a fellowship from the John Simon Guggenheim Memorial Foundation.

Finally, Susan Abrams went far beyond the duties of editor and friend by encouraging me and helping me see how to assemble my disjointed material. More than once, Susan and her coeditor, Christie Rabke, transformed the labor of writing into a rewarding social endeavor.

Thanks to you all.

Minneapolis, 3 January 1994

Credits